W. KURT
12/2019

Pennsylvania's
Amazon Princess Railroad

William Lawrence Adams

AuthorHouse™
1663 Liberty Drive
Bloomington, IN 47403
www.authorhouse.com
Phone: 1-800-839-8640

© 2013 William Lawrence Adams. All rights reserved.

No part of this book may be reproduced, stored in a retrieval system, or transmitted by any means without the written permission of the author.

Published by AuthorHouse 7/03/2013

ISBN: 978-1-4817-7354-6 (sc)
ISBN: 978-1-4817-7353-9 (hc)
ISBN: 978-1-4817-7352-2 (e)

Library of Congress Control Number: 2013912262

Any people depicted in stock imagery provided by Thinkstock are models, and such images are being used for illustrative purposes only.
Certain stock imagery © Thinkstock.

This book is printed on acid-free paper.

Because of the dynamic nature of the Internet, any web addresses or links contained in this book may have changed since publication and may no longer be valid. The views expressed in this work are solely those of the author and do not necessarily reflect the views of the publisher, and the publisher hereby disclaims any responsibility for them.

Contents

Research Note: Pennsylvania's Lost Legend Trail vii

One Final Note xiv

Preface xvi

I. The Origin of the Enterprise 1

II. Colonel George Earl Church 22

III. The English Failure and Its Consequences 28

IV. The Organization of the American Expedition 56

V. The Voyage of the Mercedita 67

VI. Barbados 83

VII. Voyage of the Mercedita Continued 90

VIII. Pará 94

IX. Pará to San Antonio 104

X. San Antonio 123

XI. The Wreck of the Metropolis 135

XII. Departure of the City of Richmond 164

XIII. Mackie, Scott, and Company, Limited 170

XIV. The Sea Voyage of the Juno and Brazil	180
XV. Events at San Antonio	214
XVI. With Byers at Macacos	234
XVII. A Preliminary Reconnaissance on the Upper Madeira	241
XVIII. Events at San Antonio Resumed	259
XIX. Two Months with Captain Stiles	272
XX. At Headquarters Again	294
XXI. The Canoe Voyage of Corps No. 1 to San Patricio	310
XXII. San Patricio to the Jaci Paraná	323
XXIII. The Search for Bruce's Corps	341
XXIV. Camp Life in a Rubber Forest	349
XXV. Our Anthropophagous Acquaintances	354
XXVI. Last Days on the Jaci Paraná	364
XXVII. Disintegration and Collapse	372
XXVIII. Mr. Mackie's Experience at Pará	403
XXIX. Results	424
XXX. The Litigation in England	432
XXXI. The Recent Revival of Colonel Church's Project	446
Appendix	456
Index	469

Research Note: Pennsylvania's Lost Legend Trail

In the year 1772, John Adams commissioned a corrected map of South America. The work was done by Bowen Sculp.

- ✓ In 1839 (a full century before this investigative reporter was born in the rolling hills of Western Pennsylvania), Emperor Don Pedro of Brazil abdicated the monarchy and named his five-year-old son, Don Pedro II, to be emperor. An education in the Greek, Latin, and Roman classics followed, and by the time Don Pedro II came of an age to rule on his own, he had adopted the liberal attitudes of North America in regard to slavery.
- ✓ In 1852, Navy Lieutenant Ladner Gibbon, a Philadelphia, Pennsylvania, native left on a mission of international intrigue. He was tasked with making the first comprehensive survey of the Upper and Lower Amazon River Complex. The survey was sponsored by Bolivia, Brazil, and the United States.
- ✓ In Brazil, in 1850, the importation of African slaves had been prohibited, but slave-breeding farms were established to supply labor for the cotton plantations.

It was not until 1871 that children of those slaves were granted their liberty.
- ✓ The major political issue of the period from the 1860s to the 1880s was slavery and the separation of the power of the church from the will of the people.
- ✓ Don Pedro and many of the liberal politicians were convinced that the institution of slavery was outdated, but the economy was dependent on the two institutions, and it was not until 1880 that the Brazilian Abolitionist Society was founded.
- ✓ In 1867, under external pressure, Brazil declared the entire Amazon River Complex (the ARC) open to navigation by all nations, thus providing Bolivia with a trading route access to the Atlantic.
- ✓ In 1869, Ulysses S. Grant became president of the United States of North America. Don Pedro II, as emperor of Brazil, found a man for whom he had great admiration. Grant extended an invitation to visit the United States.
- ✓ In 1876, in Philadelphia, the United States created an international sensation with the opening of the Philadelphia Centennial Exposition (Fair) to celebrate their independence, the end of the US Civil War, and the beginning of the industrial revolution in the Americas.
- ✓ Don Pedro thus became the first—and only foreign—head of state to visit the fledgling international economic power of the Americas.
- ✓ During his visit to North America, he addressed the US Congress and told of his intentions to one day

join the North American countries as a sister nation in the pursuit of an Americas Empire, to be known as the United States of the Americas. This, the joining of the North and the South American continents of the Western hemisphere was the inevitable outcome of the 1823 Monroe Doctrine, whereby the Americas gave fair warning to the European powers to either assimilate or vacate.

- ✓ At the Centennial Exposition, Don Pedro II, acting as the director, led the orchestra in a rendition of his own creation dedicated to a joint venture for the future of the Americas.
- ✓ When the exposition was officially opened to the public, it was Emperor Don Pedro II of Brazil and President Grant who, hand on hand, threw the switch that turned on the Corliss Engine, as the feature event that provided the electric power for the entire thirteen acres of industrial machinery.
- ✓ During his visit with Congress in Washington DC, along with some of the members of Congress and President Grant, at the exposition, he gave voice to his intentions of opening Amazon River trade routes from the border with Bolivia all the way to Philadelphia and beyond.
- ✓ At the exposition, he met many of the United States' inventors and industrial leaders. One inventor was Alexander Graham Bell, who had brought his latest invention, the telephone. Upon witnessing a demonstration, the emperor was heard to say, "My God, *it* speaks!"

- ✓ And, with his immediate investment of ten thousand dollars (in 1876 dollars), he was likely the first investor in what was to become the Bell Telephone Company. He took one back to Brazil to connect his home with his palace.
- ✓ He also took a railroad tour to Western Pennsylvania, traveling around the famous Horseshoe Curve and through the tunnels penetrating the Allegheny Mountains. During his visit, he was introduced to some of the engineers who were responsible for the construction of those systems of transportation. Impressed by their accomplishments, he related the dismal results of a recent British railroad-building expedition near the Bolivian-Brazilian border, where after a month-long drinking fiasco, the British had abandoned the area and the contract. They called the place the devil's cauldron and left behind only two huts built of thatch, one of which was filled with empty bottles.
- ✓ In the *Philadelphia Times* newspaper, a great deal of coverage of the exposition was the daily news, as was the visit of the first head of state of a foreign nation, and as chance would have it, the owners of the newspaper, brothers Thomas and Peter Collins, were also railroad builders who had been intimately involved with both the Horseshoe Curve and Allegheny Mountains Tunnels.
- ✓ One item of importance published in November 1877 concerned the expected trade with Brazil. Coffee, Brazil's most important export at the time, brought in

thirty-six thousand US dollars yearly, but US exports to Brazil amounted to only seven million dollars in back-haul freight.
- ✓ Congressman Stewart of the Third District of Minnesota introduced a bill in the House whereby Congress would authorize two hundred thousand dollars annually for the development of a mail steamship line between the United States and Brazil. The six-million-dollar contract went to the railroad builders, the Collins brothers, for the Mamore-Madeira Railroad and Freight Line.
- ✓ Great Britain and Holland, primarily, controlled all freight with South America and were not about to allow competition. During the course of the financial agreements and construction of the South American enterprise, they found the means to tie up the funds through fraud and outright legal theft, and in the end, the enterprise was bankrupted.
- ✓ Thomas and Peter Collins had been born, raised, and educated in the rolling hills of Cambria County, Pennsylvania, as were the explorers Admiral Robert E. Perry of North Pole fame and yours truly, who, over a period of sixteen years and in a very minor fashion, became a living legend in the State of Para (Pa), Brazil, with the development of the (thirteen-acre) FEMAZON State Fair of 1989.
- ✓ As a sidebar, it should be noted that the FEMAZON was designed around a central theme of developing a new wood by-products industry, as a means of better utilizing the trees harvested, and as such reduce the total raw materials used by foreign importers. One

species of tree in particular, the tower ray oak, was used as the focal point for the wood by-products displays.
- ✓ At the same time in history, back in Cambria County, Pennsylvania, there was another centennial celebration—that of the Johnstown Flood—and as part of the celebration, the Flood Museum was being completely renovated by the Berkibile Brothers Construction Company. They were provided with enough of the same species of Amazon hardwood imported from Para, Brazil, to be used wherever it could be made available for public viewing, showing its versatility.
- ✓ FEMAZON, meaning "faith in the Amazon people," was designed as a Brazilian centennial celebration of the emperor's 1889 abdication and subsequent travel to France with serious medical problems.
- ✓ Thereupon, the Brazilian Army took control of the country, and there occurred its renaming as the United States of Brazil.
- ✓ At the eastern end of the Amazon River Complex (the ARC) lies the ancient city of Belem, formally known as Para.
- ✓ Don Pedro II, being an aficionado of classical music, and seemingly, as a tribute to the people of the ARC, caused to be built their first formal opera house, the Theatro de Paz (Theater of Peace), which was completed in 1889 on what is today Presidente Vargas Avenue, when his beloved country became the United States of Brazil.

- ✓ Two years later, Emperor Don Pedro II succumbed to his illness and was buried in France.
- ✓ Finally, during the 1890s, one hundred academics, journalists, and political minds were tasked with submitting recommendations for the creation of the United States of the Americas agenda. Quite possibly, this was the source of the eventual Organization of the American States (OAS) in 1947. The OAS consisted of every country (state) of the Americas' hemisphere. In November of 2005, the presidents of the two largest nations of North and South America met in Brazilia, Brazil, to discuss the Western Hemisphere's Free Trade Association.

One Final Note

There exist many myths to be discovered about the ARC of northern Brazil. In each of the myths, there can always be found at least one tiny grain of truth, if one has the patience to listen closely and often.

One is about the Biauna, the princess of Para. She was a princess of the river people. Supposedly, she was born of a king from the north, and it is believed among the children that she will search worldwide for a mate who will bring great happiness and joy to the people. She was born (at) the Theatre de Paz or close by while the opera house was being built.

In the archives of the city of Belem (a.k.a. Para), there exists a possible grain of truth, in that there was a celebrated birth of a girl baby in 1878. She was born of parents not of the local population and was treated as royalty by her father who was a "king" from Juniata, Pennsylvania, or so the records state.

She was named Maria (Juniata) King by her father, Charles F. King, one of the engineers of the Pennsylvania Railroad Building Expedition of 1877 to 1879. He was from the Juniata River region of the eastern foothills of the famous Allegheny Mountains. His wife was one of three women who accompanied their husbands "up the River Amazon." In dedicating this book of lost history to the robust people of Pennsylvania, I chose to name the book, *The Pennsylvania Amazon Princess Railroad*.

The original port, constructed to receive equipment for the Collins' Brothers Expedition, in time, became a town and then a city. When Brazil's newest state, Rondonia, was created in 1980, Porto Velho (Old Port) became its capital city.

Semper Fi

W. L. "Bill" Adams
USMC Retired f/
Cambria County, Pennsylvania

Preface

For many years past, it has been customary for an organization known as the Madeira and Mamoré Association to hold an annual reunion at one of the leading hotels in Philadelphia.

The participants were persons who, in the year 1878, had gone to Brazil under the leadership of two firms of American contractors—P. and T. Collins and Mackie, Scott, and Co.—for the purposes of constructing a railway around the falls and rapids of the Upper Madeira River and establishing steamboat lines above and below the obstructions so as to form, in connection with ocean steamships plying between New York and Brazilian seaports, one great system of international transportation. The intention was to produce a rapid development of all that vast and fertile territory drained by the Amazon and establish direct communication between the United States and interior Bolivia.

There was nothing visionary or chimerical about the project, which had engaged the attention of members of the association. It originated in recommendations made by

Lieutenant Lardner Gibbon, United States Navy (USN), in 1852 and had been advocated by many eminent statesmen of South America; endorsed by noted German engineers; subsidized by the governments of Bolivia and Brazil; and entrusted for execution to Colonel George Earl Church, a distinguished soldier in our Civil War. It failed in consequence of legal and financial complications in England and, after careful investigation, has again been revived by the government of Brazil, which, by terms of a recent treaty with Bolivia, is pledged to its immediate execution.

The members of the Madeira and Mamoré Association had long felt that the true history of the enterprise and of their disastrous experience, while connected with it, was worthy of preservation. Few of the many newspaper accounts published at the time were correct, and none were complete; those who could speak from personal knowledge of occurrences in Brazil were fast passing away, and those still living and qualified to write the desired history were too actively engaged in the living present to occupy themselves in raking over the dead ashes of the past. At the same time, the constantly increasing importance of executing the work they had failed to perform made it probable that knowledge of their experience would prove valuable to those who, in the not-too-distant future, would inevitably follow in their footsteps.

This was the situation, when, in February 1903, the writer first found it possible to attend a reunion of those whom he had accompanied to Brazil a quarter of a century before, and ten minutes after his appearance, a formal resolution was passed imposing upon him the duty of writing the entire history of the Madeira and Mamoré enterprise.

To the present day, he has never been able to fathom the motives that induced his associates to make such an assignment, but it is highly probable that knowledge of his then recent retirement from active business life, his consequent ability to devote the requisite time to the task, and his sustained interest in South American affairs, rather than any actual or supposed literary qualifications, led to his appointment.

At each subsequent reunion, an installment of the proposed history has been read or distributed in pamphlet form, but this method of presenting the narrative in detached parts, it soon became evident, was impracticable and unsatisfactory. In the slow process of gathering facts from many different sources, he was constantly receiving information, which made it necessary to modify, amplify, and correct parts already printed, and it was impossible, they discovered, in annual readings of one hour each, to exhaust the subject during the probable lifetime of members of the association.

These considerations led to the present publication in book form, which the writer submits to his former associates and others, fully conscious of its imperfections and claiming for it no other merit than that of being a truthful statement of his own recollections and all the facts he has been able to collect, during several years of diligent research, concerning one of the most remarkable enterprises ever undertaken by Americans on foreign soil.

While our national energies are being wasted in the far-distant Philippines, this story may serve to draw attention to a legitimate field for commercial enterprise nearer home; while we properly appreciate the importance of constructing the Panama Canal, it may call to mind thousands of miles of natural waterways, where the American flag is looked upon as a rare curiosity, and during the delirium caused by unexampled national prosperity, the history of an American failure may prove to be a valuable sedative.

Thoreau has said, "If our failures are made tragic by courage, they are not different from success." That both tragedy and courage enter largely into the following narrative will not be questioned by those who read of the wreck of the *Metropolis*, of the two diminutive river tugs sent on a voyage of 3,200 miles by the direct sea route from Philadelphia to Pará, and of the 221 men who lost their lives on land and at sea in a vain endeavor to open the heart of the South American continent to the commerce of the world.

Unqualified success is not infrequently the mere logical sequence of unlimited resources, rather than a consequence of skill and training. All history proves that "Calamity is man's true touchstone." Failure and disaster quickly differentiate the degenerate and cowardly weakling from the man capable of self-sacrifice in devotion to an enterprise far more important than the personal welfare of those engaged in its prosecution. The inevitable result is promptly to terminate the career of those unequal to the struggle and to develop in others a self-reliance that more prosperous circumstances would have permitted to remain dormant.

This seems to explain the remarkable professional success that has since attended many of those engaged in the ill-fated expedition of 1878, doubtless in a great measure due to the strength of will and force of character developed in the unsuccessful attempt of American contractors to survey and construct the Madeira and Mamoré Railway, through previously unexplored primeval forests and tropical jungles, around the falls and rapids of the Madeira River in Brazil.

The writer here desires to say that the complete history of this South American enterprise could never have been written had it not been for the mine of information contained in a report made by Colonel Church to the governments of Bolivia and Brazil, entitled *Bolivia via the River Amazon*, his history of "Explorations in the Valley of the River Madeira," and numerous papers by the same author, which appear in the official publications of the Royal Geographical Society.

He desires also to acknowledge valuable assistance given him by the following persons: Mr. W. R. Taylor, vice president and secretary of the Philadelphia and Reading Railroad; Mr. Charles P. Mackie, of Englewood, New Jersey, formerly of Mackie, Scott, and Company, Limited; Mr. Othniel F. Nichols, the resident engineer at San Antonio for the Madeira and Mamoré Railway Company, since widely known as chief and consulting engineer in the Department of Bridges, New York City; Colonel John J. Jameson, of Philadelphia, who was general manager for P. and T. Collins at San Antonio pending the arrival of Mr. Thomas Collins at that place; Mr. Camille S. d'Invilliers, who, after the resignation of Mr. C. M. Bird, became chief engineer of the Madeira and Mamoré Railway and has ever since the abandonment of that undertaking been engineer of construction on the Pennsylvania Railroad at Cresson; Mr. William C. Wetherill, of Denver, Colorado, who served as topographer and principal assistant engineer in Brazil and has since held many positions of great responsibility, among them that of chief engineer of the Mexican National Railway; Mr. George W. Creighton, who filled various positions connected with the railway engineering work in 1878, even then giving marked evidence of that energy and ability that have since made him general superintendent of the main line of the Pennsylvania Railroad between Philadelphia and Pittsburgh; Mr. Cecil A. Preston, who was assistant engineer, acting principal assistant engineer, and engineer of construction in Brazil but is now superintendent of the Middle Division of the Pennsylvania Railroad at Altoona; the Honorable Charles F. King, formerly the leading subcontractor on the Madeira and Mamoré Railway, later a state senator from Schuylkill County, Pennsylvania, and, for many years past, member of a prominent firm of engineers and contractors in Philadelphia; Mr. Joseph S. Ward, the able and efficient secretary of the Madeira and Mamoré Association, who joined the engineer corps of the Madeira and Mamoré Railway during the early stages of a long and useful career, which he is still actively pursuing as resident engineer for the Philadelphia and Reading Railroad at Williamsport, Pennsylvania; Mr. Robert H. Hepburn, of Avondale, Pennsylvania, who, as acting general manager for Mackie, Scott, and Company, had charge of the two little steam tugs dispatched from Philadelphia to Pará; Mr. Charles J. Hayden, formerly assistant engineer in Brazil, but now right-of-way and tax commissioner for the Great Northern Railway at Saint Paul, Minnesota; Mr. Franklin A. Snow, who was assistant to the resident engineer at San Antonio in 1878 and is now a prominent civil engineer of Boston; Mr. John P. O'Connor, who had a most extraordinary experience while connected with the engineer corps of Captain Stiles in Brazil and is now private secretary to Archbishop Ireland at Saint

Paul; Mr. Thomas A. Shoemaker, the well-known contractor of Bellefonte, Pennsylvania, a relative of the Messrs. Collins and the possessor of many interesting papers relating to their Brazilian contract; and Professor S. C. Hartranft, of the Northern Normal and Industrial School at Aberdeen, South Dakota, who with others in a canoe descended the Madeira River in 1878 and, without an adequate supply of money or provisions, was so fortunate as to reach home alive.

I. The Origin of the Enterprise

How few think justly of the thinking few!
How many never think, who think they do!
—June Taylor

The Republic of Bolivia, as it existed prior to the termination of the war with Chile in 1882, had an area of 597,271 square miles, exclusive of the territory of El Chaco, claimed alike by Bolivia, Paraguay, and Argentina.

The population, though never carefully determined, was estimated by the best Bolivian authorities as two and a half million, and of this, about half consisted of savage and domesticated Indians.[1] In other words, a population about equal to that of the State of Massachusetts occupied a territory three and a half times greater in area than that covered by our ten New England and Middle States combined.

During the colonial days of South America, Bolivia was a part of Peru, having been subdued and annexed by Hernando, a brother of Francisco Pizarro, and in 1559, it was formed into the Audiencia of Charcas, or Upper Peru. "The haughtiest of

---

[1] Bureau of American Republics.

all the old Spanish Conquistadores," says a prominent writer, "settled in the country and clustered their titled families around its ten thousand open silver-mines."

The treasures poured into Spain from Potosi and other mines "exercised a marked influence in European history and largely furnished the sinews which enabled Spain to hurl her armies against France, the Netherlands, and Portugal, to turn back the Turks at Lepanto, and fit out the Invincible Armada against England." In mineral resources, Bolivia is probably the richest country in South America.

The former output of her mines, though well authenticated, seems almost incredible. She has rich deposits of gold, silver, copper, tin, lead, mercury, and rock salt. Beds of coal are being worked today, and iron is known to exist.

The silver deposits of Potosi alone, from the time of their discovery in 1546 to 1864, are reported to have yielded the enormous sum of $2,919,899,400.

Nor are these mineral resources by any means exhausted today. The almost impassable barriers to the transportation of machinery, the spirit of revolution, and the indolence, ignorance, and poverty of the people have all combined to render Bolivia's wealth unavailable, but that it exists today there is abundant proof.

The correspondent of a prominent American newspaper writing from La Paz, Bolivia, only a short time ago, says:

> The ores of Potosi, Cerro del Pasco, and other noted mines are so rich that a yield of two hundred dollars per ton is common, even with the primitive methods now employed. Remembering that miners of the United States find it profitable to work mineral worth

only ten dollars per ton, one may form some idea of what these Andean treasure-houses might be made to disclose in the hands of wide-awake workmen aided by modern machinery. The gold and silver fields of these countries are producing comparatively little nowadays, owing to the poverty and ignorance of the people, their poor methods and worse management. For many years, numbers of the mines most famous in history have had nothing done to them beyond the reduction by modern processes of the refuse of the ancient miners. ... Gold has been found in many places, but has never been extensively mined, being much harder to get at by processes in vogue while silver is so plentiful that the people can afford to dispense with the more precious but troublesome metal. To this day big nuggets of pure gold are occasionally picked up by some wandering prospector. ... Years ago Potosi received its greatest boom by means of a stroke of lightning, which detached a mass of solid gold from some unknown cliff away up the mountainside and dropped it at the feet of a group of miners in the vale below. For a long time this mysterious nugget was the wonder of the world; then it was sold at a fabulous price to the Royal Museum at Madrid, where it may still be seen. ... If the spirit of revolution ever remains "laid" long enough for capitalists to feel secure in investing their money here, and if roads are constructed so that the products of interior Bolivia may find an easy outlet to the sea and proper machinery for working the mines find ingress by the same means, a renaissance may occur which will remind the world of the El Dorado of olden times.

In recent years, there has been a marked increase in the productiveness of Bolivian mines, and one of the latest works on Bolivian history[2] contains this statement:

> The production of silver rapidly increased, reaching fifteen million dollars in 1885, when Pacheco was president, and growing to twenty millions in 1888 with Arce in the executive chair. Potosi still yields three million ounces per annum, and the great Huanchaca mines far surpass Potosi, making Bolivia the third silver producing country in the world. But her resources can never be profitably utilized until a practical outlet to the sea has been found.

Rich as Bolivia undoubtedly is in minerals, these constitute but a small part of her natural resources. She is possessed of a large area of remarkably fertile land well watered by noble rivers and a climate that has almost every possible variation between the snowcapped peaks of the Andes and the tropical heat of her lowlands. Within her own borders, she produces nearly all the cereals, vegetables, fruits, and meats one usually expects to find anywhere in the temperate or torrid zone. The animals required by present methods of transportation or necessary to furnish subsistence and wool for clothing are raised almost without care or expense. Her forests furnish game, and her rivers teem with fish.

Bolivian tobacco is incomparably superior to any other.

---

[2] Thomas C. Dawson, *South American Republics*.

India rubber, cotton, sugarcane, cinchona bark, cacao, vanilla beans, the coca plant, gums, dye-woods, various medicinal plants, and many kinds of ornamental wood, highly prized by the cabinetmaker, are familiar products of her soil. The commerce of Bolivia is, however, insignificant in comparison with what it might be with better facilities for reaching the seaboard. At the time of which we write, she had but one seaport, which was comparatively useless, because it was situated in a waterless desert that she has since surrendered to Chile. Almost her entire commerce had been for many years conducted through this port of Cobija in the Desert of Atacama and the port of Arica in Peru. These seaports could be reached from interior points in Bolivia only after a vexatious trip over the Andes at an elevation of fifteen thousand feet by paths along precipices where even the sure-footed mule at times found it difficult to go and where the penetrating cold and the *soroche* made life almost unendurable to those accustomed to lower altitudes and a warmer climate. The cargo that could be conveyed by this route was limited to the carrying capacity of a mule, which was never more than three hundred pounds; on the eastern side of the Andes, it was limited to one hundred and fifty. It can easily be seen that the transportation of bulky or heavy articles, such as machinery, by any such method was out of the question.

In the year 1870, a railway of standard gauge, 325 miles in length, was opened for traffic between Porto Mollendo and Puño freight, and passengers were then conveyed by steamer over Lake Titicaca to Chililaya, 110 miles, and thence by a good and comparatively level wagon road forty-five miles, to La Paz in Bolivia. In the year 1892, a narrow-gauge (seventy-five centimeters) railway was completed between the Chilean port of Antofagasta and Oruro in Bolivia, a distance of 573 miles, but, owing to heavy grades and sharp curves, trains only ran over it in daylight and it required three days to travel over the entire route.

By the treaty between Chile and Bolivia, signed October 21, 1904, the government of Chile guarantees the completion of the railway between Arica and La Paz.

Several other railways are projected, but at the present time, they exist only on paper. These railways have, to a great extent, diminished the difficulties formerly encountered on the Pacific side of the Andes, but it must be remembered that La Paz, Puño, and Oruro are all situated in the Titicaca Basin and all have elevations of more than twelve thousand feet above sea level. The difficulties involved in extending any of these railways to the lowlands of Bolivia on the eastern side of the Andes would be so enormous that no traffic, present or prospective, would justify us in regarding them as possible competitors for the trade that naturally belongs to the great waterways with which Bolivia is so abundantly provided.

The interior railways of Bolivia will have an important place in the future development of the country, but they will be local in character and serve mainly as adjuncts to steam navigation on the rivers. It is, therefore, quite as true today as it ever was that the route over the Andes to the Pacific coast does not furnish an adequate outlet for the commerce of Northern and Eastern Bolivia.

Some attempts have been made to discover a satisfactory highway for Bolivian commerce by way of the Paraguay River and the Rio de La Plata. Two railways have been projected to connect Oruro, Cochabamba, and Sucre with navigable water on the Paraguay. The known facts in regard to this route, while not conclusive, are far from encouraging.

Colonel George Earl Church, in a recently published pamphlet, has ably discussed this subject.

He admits that vessels drawing nine feet of water may reach Asuncion, the capital of Paraguay, and that vessels drawing three feet may at all times ascend the Paraguay River to the mouth of the Rio São Lourenço. In order to reach navigable water on the Paraguay from any point on the Andean foothills, it is, however, necessary to cross the bed of the ancient Pampean Sea that at one time extended on both sides of the Paraguay River and as far north as the mouth of the Mamoré. The vast extent of this great basin and the difficulty of constructing a road of any kind through it will be more thoroughly appreciated after reading the following statement. Referring to the Mojos Basin, Colonel Church says: "During a period of about four months of the year, some 35,000 square miles of its surface are covered with the surplus water that cannot find exit over the falls of the Madeira."

In June 1852, when the floor waters were disappearing, Lieutenant Gibbon, of the United States Navy, made a trip from Trinidad in the Mojos Basin to Loreto, a distance of twelve leagues, and he thus speaks of the country along his route:

> The roads are navigable for canoes half of the year, when traveling is much more easy than when the season is called dry. The Indian builds his hut on those elevated places which remain islands; when the great flood

waters come down, crickets, lizards, and snakes crawl into his thatched roof; droves of wild cattle surround his habitation, armadillos rub their armor against the pottery in the corner of his hut, while the tiger and the stag stand tamely by. The alligator comes sociably up, when the "gran bestia" (the tapir) seats himself on the steps by the door.

The animal family congregate thus strangely together under the influence of the annual deluge. Those of dry land meet where the amphibious are forced to go, and, as the rains pour down, they patiently wait. Birds fly in and light upon the trees and top of the hut, while fish rise from out the rivers and explore the prairie lands.

Colonel Church quotes Count Castlenau as saying, "All the plains from the mouth of the Mamoré to the Pilcomayo are inundated from October to March, and present the aspect of a great ocean dotted with green islands." In another place, Colonel Church, speaking of the Pilcomayo River, which traverses the bed of the ancient sea, tells us that "From 180 to 200 miles above its mouth, the Pilcomayo filters through a vast swamp about 100 miles in diameter, through which there is no principal channel."

Again the same author states:

> From the junction of the São Lourenço (or Cuyaba) with the Paraguay, the latter, now a great river, moves sluggishly southward, spreading its waters in the rainy season for hundreds of miles to the right and left as far as latitude 20° south, turning vast swamps into great lakes like that of Mandioré on the Bolivian side; in

fact, temporarily restoring the region for thousands of square miles to its former lacustrine condition.

Reviewing an account of explorations on the upper Paraguay by Captain Henry Bolland, Colonel Church says, "All attempts to find a practicable route across the Bolivian Chaco from the Paraguay River south of latitude 19° south have been failures; and, according to the season, the traveler who attempts it may die of thirst or be drowned before he reaches the foot hills of the Andes 400 or 450 miles distant."

In another part of the same review, regarding the route from Asuncion to Santa Cruz, we are told that "In the dry season the entire 570 miles is as wanting in water as it is overflowing in the wet one, and oxen and beasts of burden frequently died of thirst in attempting the journey." Altogether, the prospect of establishing railway communication between interior Bolivia and the head of navigation on the Paraguay is not such as to excite enthusiasm. The exploration and development of any such route has been very much retarded in the past by the conflicting claims of Bolivia, Brazil, Paraguay, and Argentina in regard to their boundary lines in the vicinity of the Paraguay River. The significance of such disputes lies in the fact that nearly all the South American states at one time claimed the exclusive right to navigate all rivers within their respective territories, and it was not until September 7, 1867—and then under external pressure—that Brazil declared the Amazon open to navigation by the ships of all nations.

The Argentine railway system has of late years been pushing northward from Buenos Aires along the eastern side of the Andes and, it is possible, may in time afford an outlet for some part of the commerce of Bolivia, but the route, either by the Paraguay River or by the Argentine railways, would be a very circuitous one for reaching any port in Europe or the United States.

From the year 1541, when Orellana made his famous descent of the Amazon in search of the fabled El Dorado, a period of two centuries elapsed before the advantages offered by the Madeira and Mamoré Rivers, as a means of internal communication, began to be faintly appreciated. In the year 1716, an exploring expedition ascended the lower Madeira. In 1719, the Captain General of Pará was killed by a falling cedar on the Madeira, and the Portuguese were extending their settlements up the principal tributaries of the middle Amazon.[3] Southey tells us that:

> The Madeira had been navigated before this time (1742). It is said that so early as the days of Nuflo de Chaves, when the first settlement of Santa Cruz was abandoned, a party of the more adventurous inhabitants went along the Moxos tribes and, embarking in their country, either upon the Ubay or the Mamoré, followed the stream boldly as Orellana and with like good fortune till they reached the main sea.

He tells us further that "A Carmelite also had reached Exaltacion; he had ascended the river from the most advanced of the Pará missions."

In the year 1723, the governor of the state of Pará, hearing from men who had been on the Madeira that Europeans

---
[3]

had settled above the falls of that river, sent a troop of men, commanded by Francisco de Mello Palheta, to explore and report upon the country and to ascertain the nationality of the settlers. After reaching Exaltacion in Bolivia, the expedition returned to Pará and made the earliest voyage, of which we have any record, between interior Bolivia and the Atlantic Coast by way of the Amazon and tributary streams.[4]

The mines of Mato Grosso, which had been discovered in 1734, drew many adventurers overland from the Portuguese settlement at São Paulo. One of these, Manoel Felix de Lima and a party of men in 1742 made their way in canoes down the Sarare River to its junction with the Guaporé and followed this river to the mouth of the Baure, which they ascended to the Mission of São Miguel. Later, they reentered the Guaporé and continued down that stream until they reached the Magdalena, which they ascended to the Mission of São Maria Magdalena. From this point, the majority of the party set out for Exaltacion, but Manoel Felix de Lima, with three companions, continued down the Magdalena to the Guaporé, or Itenez, River, which they followed to its junction with the Mamoré. Continuing down the Mamoré, Madeira, and Amazon, after many trials and adventures, they finally reached Pará. This voyage was remarkable as it was the first to open up river communication between Mato Grosso and Pará.[5]

In the year 1749, by special order of His Faithful Majesty, the King of Portugal, José Gonsalves da Fonseca

---

[4] Colonel George Earl Church, "Explorations in the Valley of the River Madeira," *Col. George Earl Church*.

[5] Southey, *History of Brazil*.

made a voyage in canoes from the city of Pará, up the Amazon, Madeira, Mamoré, and Itenez Rivers, to reach the mines of Matto Grosso, which, though successfully accomplished, occupied nine months and was apparently so fraught with dangers and vicissitudes as to discourage frequent repetition.

Nevertheless, Southey, in commenting on it, says:

> From that time the navigation between Mato Grosso and Pará was frequented, notwithstanding the length and difficulty and danger of the way. It was found that Mato Grosso could be supplied at a cheaper rate with European goods from Pará than from Rio, and the voyage was far less perilous than from S. Paulo.

Thaddeus Haénke, a member of the Academy of Sciences of Vienna and Prague, and commissioned by His Catholic Majesty to explore Peru, writing in 1799, deplored the jealousies then existing between the Portuguese and the Spaniards and pointed out the vast treasures in Bolivia and their value to the world. He offered to undertake the opening of navigation down the Mamoré, the Madeira, and the Amazon and enumerated the rare products that would furnish cargoes for his vessels, but a war of independence had to be fought out and republics and a monarchy had to be established in South America.

Señor José Augustin Palacios, a Bolivian engineer, in 1846 ascended and descended the Madeira and Mamoré Rivers in a canoe, though compelled to make portages at the falls and rapids. The following statements in his published report deserve reproduction here:

> This inconvenience (the passage of the falls of the Madeira) is one that might easily be surmounted. ... Our statesmen, instead of thinking about Arica or Cobija, should direct all their energies and attention to the navigation of the Madeira. ... The project of emigration, as well as that of navigation, unfortunately, cannot be carried out until the cessation of the dissensions and internal feuds which distract the attention of the governing authorities from an object of such great importance, which is to place the whole of oriental Bolivia in contact with the great markets of the United States of North America, whence their commodities would be imported with considerable utility. Under this influence the sciences, commerce, arts, and all of the useful manufactures would advance. I pray Heaven that this wished-for day may arrive, when beneath the shadow of peace Bolivia may become the precious gem of the continent and add new glory to the immortal genius of her founder.[6]

Writing from Manáos, Brazil, under the date of February 24, 1868, Señor Ignacio Arauz, an energetic and enterprising Bolivian, who had a thorough practical knowledge of the obstacles to the commercial development of his country, said:

> On the one hand, while Europe cultivates ideas of exalted progress, facilitates means of communication in all directions as far as her most remote confines,

---

[6] Colonel George Earl Church, "Explorations in the Valley of the River Madeira."

while her ingenuity invents means of convenience for the most trifling wishes, and tills her almost sterile soil, so as to produce more than might be expected from the nature of the locality; on the other hand, the people in the eastern part of Bolivia, exuberantly rich though it is in all the kingdoms of nature, are obliged, in order to learn the complicated machinery of a steamer and the simple apparatus of a sail, to cross the mountain chain of the Andes, and traverse the wide and arid desert of Atacama, exposed to succumb through thirst, and suffering the rigor of inclement weather, in a march of more than 320 leagues, as far as Cobija. This is the real obstacle to the progress of Bolivia.[7]

To a native of Philadelphia, Lieutenant Lardner Gibbon of the United States Navy, who is still living, is due the credit of making the first thorough exploration of the route from Bolivia to the sea coast at Pará by way of the Chaparé, Mamoré, Madeira, and Amazon Rivers. In the year 1851, Lieutenants Herndon and Gibbon were ordered by the secretary of the navy to proceed overland from the coast of Peru and examine the more important waterways connecting the interior of Bolivia and Brazil with the seaboard. These two young officers separated at Tarma in Peru, took different routes over the Andes, and followed different tributaries leading into the Amazon. Lieutenant Gibbon went overland to Vinchuta, in Bolivia, and thence by canoe all the way to Pará. Notwithstanding the fact that his party was small and poorly equipped, one cannot read his report today without a

---

[7] Ibid.

feeling of astonishment at the amount and reliability of the information he was able to collect in a very short time. He corrected the best existing maps by his own astronomical observations, made careful report on the resources of the country through which he passed, took soundings in the rivers, and made many valuable notes regarding navigations, climate, and temperature. An accident having destroyed his barometer, with no other instrument than a hastily improvised boiling-point apparatus, consisting of a thermometer and a coffeepot, he took observations for altitude that enabled him to construct an approximately correct profile of his entire route across the continent of South America.

He found that a series of nineteen falls and rapids, extending from Guajará-merim to San Antonio, was the sole obstacle to continuous river navigation from Vinchuta, in Bolivia, by way of the Chaparé, Mamoré, Madeira, and Amazon Rivers to the seaport of Pará, a distance of about 2,240 miles.

To pass this obstruction, he recommended the building of a mule road between San Antonio and Guajará-merim on the east side of the Madeira and Mamoré Rivers, through the territory of Brazil, and over substantially the same route afterward adopted for the Madeira and Mamoré Railway. He estimated that by cutting across a bend in the river, the length of the proposed road would not exceed one hundred and eighty miles and proved conclusively that, by constructing this road and establishing steam navigation on the rivers, goods could be transported from the United States and Europe to the heart of Bolivia that could not be conveyed at all by any other known route. He pointed out the fact that these improvements would shorten the time of transit between Baltimore and La Paz, the commercial emporium of Bolivia, to fifty-nine days, while it required one hundred and eighteen days, under the most favorable circumstances, to make the same trip by the usual route around Cape Horn and over the Andes. He gives us a graphic picture of the difficulties involved in maintaining canoe navigation through the obstructed portion of the Madeira and Mamoré Rivers and cites the case of a Brazilian merchant who spent five months conveying a cargo in canoes upstream between the termini of his proposed mule road. Incidentally, Lieutenant Gibbon experienced a foretaste of the joys in store for those who were to follow after him and attempt to execute his project. Bearing in mind that the really unhealthy part of the river is limited to about two hundred and forty miles between Guajará-merim and San Antonio, through which he passed with comparative ease *downstream* in twelve days, recollecting, too, that time is required to develop the initial outbreak of malarial fever even in the worst climate, and, further, the well-known fact that on both North American and South American rivers, men who live all the time on boats enjoy a comparative immunity from malarial fevers not granted to those on shore, Lieutenant Gibbons's experience is highly significant. On September 19, 1852, he reached the southern terminus of his projected road, so far as his narrative shows, without having incurred any serious physical trouble beyond a slight attack of fever on the Andes and the fatigue naturally incident to such a journey. Eight days later, on September 27, at the Falls of Girão, he notes that his men "complained of headaches and pains in their backs; the strongest were jaded." At the Falls of Theotonio, on September 30, he adds: "I was attacked with a severe bilious fever which brought me at once to my back. The pain in my left breast was somewhat like that described by those who have suffered with the Chagres fever. We were all worn out and haggard."

At Tamandua Island, below San Antonio, on October 2, he remarks, "The fever kept me in bed in the canoe with pains that forbade sleep at night."

At Crato, on October 6, he tells us, "There is no rest with a high fever."

At Porto de Mataura, on October 12, he gives us the further information that "the suffering from fever was increased to agony."

At Borba, on the Madeira, a short distance above the mouth of that river, on October 14, he mentions that his host, Captain Diogo, "disapproved of sleeping, which was all we wanted, except to get out of the country as soon as possible."

The existing state of geographical knowledge in regard to the Madeira and Mamoré Rivers at the time Gibbon made his exploration may be inferred from the following absurd passage in Herndon's report, published before Gibbon's return. Herndon says: "Mr. Clay, our *chargé* at Lima, was told that a *Brazilian schooner-of-war had ascended the Madeira above the rapids and fired a salute at Exaltacion, which is in Bolivia above the junction of the Beni.*"

In 1861, General Quintin Quevedo, of Bolivia, after descending the rapids, suggested their canalization, or the establishment of steam-navigation on the rivers above in conjunction with a railroad about the falls. In the year 1863, the distinguished Bolivian Don Rafael Bustillo wrote:

Bolivia is situated on the masses of silver on the table ranges of the Andes. She has a territory fertile beyond measure, where the treasures of the most opposite climates are grouped together.

With all this Bolivia perishes for want of methods of communication which may carry to the nations of the world her valuable products and stimulate her sons to labor and industry, but these communications do not exist, nor can they exist, except in the deep-flowing rivers with which the finger of God has been pleased to furrow her soil. They are not found, nor yet can they be found except in the magnificent affluents of the Amazon.

In 1867–8, Franz Keller, a German engineer, accompanied by his brother, and commissioned by Brazil, surveyed the obstructed portion of the Mamoré and Madeira Rivers, with a view to their improvement. He estimated that inclined planes could be constructed around all the larger rapids to track vessels, as in North America and Prussia, for $450,000; that the river could be canalized with locks for $10,500,000; that a railroad could be built around them for $4,250,000; or that a twenty-foot-wide macadamized road could be constructed for $2,850,000. Keller collected a vast amount of valuable information in regard to this region and confirmed in a remarkable way the views previously expressed by Lieutenant Gibbon. He made large-scale maps of the river and took careful soundings, but, owing to the fact that his chronometers proved unreliable by reason of rough usage and that he was, in consequence, compelled to resort to lunar distances for his longitude, his maps subsequently proved to be very much distorted in an east and west direction.

In regard to the difficulties and dangers of canoe navigation through the falls of the Madeira, Keller says:

> Of the misery and annoyance of such repeated unloading and carrying of heavy chests over glowing bare rocks, under the burning rays of the sun, against which the stunted growth of the stony soil offers no shelter worth mentioning to the poor Indians, only he can form an idea who has seen this kind of *navigation* with his own eyes. Notwithstanding all this, packages of from five hundred to six hundred pounds are sometimes transported to Bolivia in the same covers in which they came to Pará; and I was told that even pianos have been thus conveyed, and—wonderful to relate—have arrived entire at Santa Cruz de la Sierra. Great as are these difficulties, they are as nothing compared with those of a transport over the Cordillera.

Keller gave a picturesque description of the savage Indians, some of them cannibals; the wild beasts; and the pestilent insects he found along his route. While he evidently sought to minimize the unhealthiness of the country, he mentions the fact that a Bolivian merchant passing through the falls of the Madeira within a few days had to bury eight of his crew, who had been attacked by fever, the rest having had a narrow escape. In regard to his own trip down the river from Exaltacion, in Bolivia, to San Antonio, between October 19 and November 18, he tells us that "All suffered in various degrees from fits of fever; which, though subdued, were not cured by repeated doses of quinine, so long as we continued to be exposed to the same pernicious influences."

Such, in brief, were the historical facts, geographical considerations, and commercial necessities that led to the definite projects for surmounting the obstructions of the Madeira and Mamoré Rivers proposed by the German engineer Franz Keller. In view of the political and financial condition of Bolivia, the ignorance of the outside world in regard to her natural resources, the fact that the obstructions were mainly in the empire of Brazil, and the great distance of the scene of operations from civilization, the execution of any one of these projects, it was evident, involved Titanic labors and demanded the services of a man possessed of a rare combination of qualities. He must be familiar with South America, its languages, its history, and its people. He must be a civil engineer of great technical and executive ability. The negotiations with the two governments immediately interested in the enterprise required that he should be a gentleman of high social standing, and, in order that his representations might carry weight in the great financial centers of the world, it was essential that he be well-known as a man of high personal character and unflinching integrity. Extraordinary as were these requirements, the Bolivian government was fortunate in being able to secure, for the work of organization, the services of a man fully prepared to meet all the exigencies of the situation. This remarkable person, whom we have several times had occasion to mention previously, was Colonel George Earl Church, whose name is today familiar to all persons of intelligence and education from Panama to Patagonia and whose life, for nearly half a century, has been largely devoted to a study of the physical geography and commercial development of South America. The personality of Colonel Church is so inseparably connected with the effort to establish a fluvial outlet for the commerce of Bolivia that the writer feels it necessary to make a slight digression from the thread of his narrative in order to give a brief and confessedly imperfect outline of his career.

II. Colonel George Earl Church

For not to any race of any clime
Is the completed sphere of life revealed;
He who would make his own that round sublime,
Must pitch his tent on many a distant field.
—Bernard Taylor

Colonel Church was born at New Bedford, Massachusetts, December 7, 1835, and those who believe in the influence of heredity on character will not have far to seek for an explanation of the sterling integrity, love of adventure, military tastes, migratory disposition, executive ability, and penchant for railway construction that became prominent characteristics in his years of maturity. In the male line, he was directly descended from Richard Church, who, in 1632, came to Plymouth, Massachusetts, from Oxford, England, and married Elizabeth Warren, whose father came to this country on the *Mayflower* and was an ancestor of General Warren, who fell at Bunker Hill.

One of the sons of Richard Church was Captain Benjamin Church, the famous colonial leader against the Indians during King Philip's War, whose heroic exploits are matters of history and who, between 1689 and 1704, commanded five expeditions against the French and Indians of Maine and New Hampshire.

On the maternal side, Colonel Church is a lineal descendant of Mary Clapp, whose maiden name was Winslow. She was a daughter of Edward Winslow, who came to Plymouth on the *Mayflower* and was three times elected governor of Plymouth Colony. Through his mother, Colonel Church is also directly connected with the Pease family of Yorkshire, England, well-known for having built the first steam railroad in England, with George Stephenson as chief engineer.

The immediate ancestors of Colonel Church moved to Rochester in 1725, and there, by grant and purchase, became possessed of about five hundred acres of land within sixteen miles of Plymouth Rock.

The greater part of this old homestead, carved out of the wilderness along the line of Indian trails, is still in possession of the family. Colonel Church's father died while he was quite young, and while he was in his eighth year, his mother moved from Rochester to Providence, where young Church attended the public schools. At the age of fourteen, he held high rank in the senior class at the high school. In his seventeenth year, he decided upon civil and topographical engineering as a profession and obtained a position on a New Jersey railroad. He was soon after transferred to one of the railroads west of the Mississippi as an assistant engineer. Later, he was employed as resident engineer at the Hoosac Tunnel and subsequently served on a railroad in Iowa.

The financial crisis of 1857 left the young man without employment, and he accepted a position as chief engineer of one of the Argentine railways and proceeded to Buenos Aires. On his arrival there, he found the country in a disturbed condition and work on the railway stopped but was almost immediately appointed to a scientific commission of military and topographical engineers ordered to explore the southwestern frontier of Argentina and report upon the best system of defense against the fierce inroads of the Patagonians and other savages living upon the pampas and Andean foothills.

The members of the commission had a most exciting experience while engaged in this work; in nine months, they rode more than seven thousand miles, and, with a covering force of four hundred cavalry, fought two severe battles with the savages. One fight, which occurred on May 19, 1859, was caused by a night attack on the commission by 1,500 picked warriors belonging to six different tribes. It was a complete surprise. Naked and mounted bareback upon their splendid horses, with their long lances in line, the savages bore down upon the expedition in a magnificent charge by moonlight. For three hours, it was a hand-to-hand fight, in which no quarter was given or asked. The savages were finally forced to withdraw, but they retired in good order, carrying with them three thousand head of cattle and horses as the fruits of their daring raid and leaving the commission in a starving condition. Each member of the commission was required to submit his own plan for the defense of the frontier, and although Mr. Church was the youngest and least experienced of the party, his plan was adopted.

In 1860, Mr. Church located the railway between Buenos Aires and San Fernando in Argentina and continued in the lucrative practice of his profession until the Civil War broke out in this country. At the first news of the impending conflict, he hastened home and promptly tendered his services in defense of the Union. During the Civil War, he served successively as captain, lieutenant-colonel, colonel, and brigade commander of volunteers in the Army of the Potomac.

At the close of the Rebellion, Colonel Church went to Mexico as a correspondent of the *New York Herald* and served in the two last campaigns against Maximilian (1866–1867). In a biographical sketch that appeared in a Boston newspaper some years ago, it is stated that, although Colonel Church went to Mexico as a private citizen and ostensibly as a newspaper correspondent, he was really on a confidential mission for General Grant and that he joined the forces under President Juarez at Chihuahua and planned the final campaign that resulted in Maximilian's capture, but that, anticipating the fate that awaited the misguided emperor, he set out for the United States, traveled six hundred miles overland in six days, crossed the Gulf of Mexico in a small steamer that almost foundered during the stormy passage, and by his representations at Washington, brought to bear the influence of this government, which he vainly hoped would save Maximilian from his impending doom.

On his return from Mexico, Colonel Church was for a time employed on the editorial staff of the *New York Herald* and later returned to South America, where he participated in some of the stirring scenes that preceded the fall of Lopez, the Paraguayan dictator.

It is unnecessary to mention here the great work done by Colonel Church in connection with Bolivian and Brazilian exploration and development, as it will fully appear in the following chapters.

In 1880, he was United States commissioner to visit and report upon Ecuador. In 1891, he represented the American Society of Civil Engineers at the London Congress of Hygiene and Demography, and in 1895, he made a voyage to Costa Rica to settle the foreign debt and to report upon the Costa Rica railway. In 1898, he became president of the Geographical Section of the British Association and in recent years has been actively interested in the projected Trans-Canada railway, intended to connect Quebec with Port Simpson on the Pacific coast by a line 248 miles shorter than that to Vancouver by the Canadian Pacific. He is now a member of the Council of the Hakluyt Society, fellow of the Royal Historical Society, vice president of the Royal Geographical Society, member of the American Society of Civil Engineers, companion of the First Class of the Military Order of the Loyal Legion of the United States, and member of the Army and Navy Club of New York City and of the Geographical, Savage, and Ranelagh Clubs of London.

Colonel Church has been a prolific writer on South American exploration and commercial development, as well as on Mexican revolutionary history. The high esteem in which he is held by the Royal Geographical Society may be inferred from a discussion of one of Colonel Church's papers read before the society in 1901, when a member said:

> We have all listened to the paper with a feeling of admiration and wonder at the immense amount of knowledge which Colonel Church possesses of the

South American continent. There is not a mountain, there is not a river, there is not a plain, there is not a portion of the sea coast, and there is not an estuary in the South American continent, of which Colonel Church is not prepared to give us full particulars.

III. The English Failure and Its Consequences

But what if I fail of my purpose here?
It is but to keep the nerves at strain,
To dry one's eyes and laugh at a fall,
And baffled, get up and begin again.
— Browning

As an essential preliminary to attempted execution of the plans proposed, a Treaty of Friendship, Limits, Navigation, Commerce, and Extradition was concluded between Brazil and Bolivia on March 27, 1867. In the same year, Bolivia had accredited a legation to Mexico under General Quintin Quevedo, who was instructed, at the termination of his labors in that country, to proceed to New York and there endeavor to find an engineer of ability willing to undertake the opening of the route to Bolivia via the Madeira and Mamoré.

On arrival in New York, the legation presented letters of introduction to Colonel Church from President Juarez of Mexico and others, at the same time inviting him to undertake the organization of the enterprise in question.

Some preliminary agreement was reached, and in 1868, Colonel Church proceeded via England to Buenos Aires and thence overland to La Paz, the Bolivian capital. There, he met the president and other officers of the government, and on August 27, 1868, he obtained a concession and made an arrangement by the terms of which Colonel Church was to organize the National Bolivian Navigation Company with a view to canalizing the falls of the Madeira and Mamoré and establishing steam navigation on the rivers above. Among other privileges, this company was given the exclusive right to navigate the Bolivian affluents of the Madeira for twenty-five years, to collect tolls from vessels not belonging to the company, which might make use of the canal, and to collect charges for carrying the freight and passengers. The Bolivian government agreed to pay the company ten thousand dollars in gold "the day the first steamer moved on the waters of the Mamoré" and made extensive land grants to the company at various points to be selected along the route of the proposed works.

Subsequent inquiry, both in the United States and Europe, convinced Colonel Church that no financial scheme that could be devised was likely to prove successful unless the Bolivian government manifested its faith in the enterprise by a guarantee of the bonds it was proposed to issue in order to raise funds with which to carry on the work. The celebrated English contractors, Thomas Brassey and William Wheelwright, had expressed their willingness to undertake the work, if such a guarantee could be secured.

Colonel Church, therefore, returned to Bolivia, and,

on November 7, 1869, some desired modifications in the original concession were made. An article was inserted in the modified concession authorizing the construction of a railway around the falls and rapids, should examination prove that method of dealing with the problem more feasible than the construction of a canal as originally contemplated. On the twenty-second of the same month, a loan contract was drawn up and executed, which provided for the placing on the European market of a loan of one and a half to two million pounds sterling, bearing interest not to exceed 8 percent. The principal and interest of this loan were guaranteed by the Republic of Bolivia, and the entire net earnings of the National Bolivian Navigation Company were pledged as additional security. An article was incorporated in the contract authorizing the issue of a supplementary loan, to the amount of five hundred thousand pounds sterling, in case the obstacles encountered proved more serious than anticipated and the proceeds of the first loan, therefore, insufficient.

The Bolivian government named Colonel Church, without solicitation or expectation on his part, as its authorized agent to negotiate the loan.

As the route of the proposed railway was entirely within the territory of Brazil, any arrangements made would be worthless without a concession from that government. Bolivia had attempted to reach some understanding with Brazil in regard to this subject but had failed. It now commissioned Colonel Church to make a second attempt. He, therefore, proceeded to Rio de Janeiro, obtained an audience with the emperor and other officials, and successfully accomplished his mission. The government of Brazil, fearing possible friction with Bolivia regarding matters entirely within Brazilian territory, preferred, however, to grant the concession to Colonel Church personally, rather than to the National Bolivian Navigation Company.

This concession was finally published by official decree of April 20, 1870, and required the organization of a company separate from the National Bolivian Navigation Company. It provided that "the enterprise shall be called the *Madeira and Mamoré Railway*" and granted to Colonel Church "the exclusive right, for a term of fifty years, to construct, pay for and possess" said railway extending from San Antonio to Guajará-merim. It stipulated that the work should be commenced within two years and be completed within seven but made provision for an extension of time, in case of delay from causes beyond the control of the concessionaire.

It gave the company, in addition to its right-of-way, certain mining and other privileges, with a land grant of thirty-two square kilometric leagues in alternate sections along the line of the railway. A kilometric league being 6.6 kilometers, or about 4.1 miles, the entire grant amounted to approximately 538 square miles.

The day following the publication of this decree, Colonel Church sailed for New York, where, on June 30, 1870, the National Bolivian Navigation Company was organized under an act of Congress passed June 25, 1870. The incorporators were:

George Earl Church	President
S. L. M. Barlow	Vice President

Directors
William H. Reynolds
Jerome B. Chaffe
James S. Mackie
Charles A. Lambard
George F. Wilson

Samuel G. Arnold and General Quintin Quevedo were afterward added to the board of directors, the latter in accordance with the terms of the Bolivian concession.

The nominal share capital was fixed at 2.5 million dollars. The Bolivian concession was transferred by Colonel Church to the company, and in return, he received two million dollars in stock. This stock was subsequently divided among the officers of the company and promoters of the enterprise substantially as follows:

George Earl Church	$1,120,000.
S. L. M. Barlow	$20,000
William H. Reynolds	$400,000
Jerome B. Chaffe	$10,000
James S. Mackie	$10,000
Charles A. Lambard	$10,000
George F. Wilson	$10,000
Samuel G. Arnold	$10,000
Morton Fisher	$400,000
An unknown person	$10,000
	$2,000,000

Colonel Church's salary, as president of the company, was fixed at £2,500 per annum. Out of the remaining $500,000 in capital stock, Julius Beer, Esq., of London, received $50,000 as a commission upon the Bolivian loan, and, in 1874, the company sold $430,000 in stock for £12,500 cash, in order to raise money for necessary expenses. This £12,500 seems to have been the only cash received by the company in return for the stock disposed of, but all of the recipients named had rendered valuable service in organizing the company, securing the necessary legislation, and furnishing legal advice. Some of them had incurred heavy personal expense in connection with the organization and had devoted much time to the work.

It will be noticed here that, as under the loan contract, all the earnings of the company were pledged as security for the contemplated loan, there was no possibility of any dividend to the stockholders until that loan had been extinguished. The stock could, therefore, have no value except that based upon faith in the ultimate financial success of the enterprise and merely represented an interest in the profits properly belonging to the company after the entire cost of the projected improvements had been paid for out of the company's earnings.

The concession expressly stated that the company should have "a *nominal* capital of one million dollars in gold, which may be augmented according to the demands of the enterprise," and did not require that any specified amount of money should be invested by the stockholders.

The prospect of dividends was too remote to induce any sane man to pay more than a nominal price for such stock prior to the negotiation of the proposed loan and application of the proceeds, in good faith, to the purposes for which they were pledged.

As the right of Colonel Church to practically sell a concession granted to the National Bolivian Navigation Company was afterwards disputed, it may be well to say here that his action was in accordance with a practice common throughout all Spanish-American countries. While it is true that the concession was granted to the company and not to Colonel Church personally, the authority to organize the company was vested in him alone. What he actually sold was not the concession itself, but the rights he undoubtedly possessed, to organize a company with the privileges and for the purposes specified in that concession. As the two million dollars in stock he received represented only possible ultimate profits, which undoubtedly belonged to the company, it would seem that the right of the company to dispose of this stock in any way could not be questioned. Neither Colonel Church nor his associates contemplated engaging in an eleemosynary enterprise for the benefit of Bolivia, and in no other way could he have secured compensation for many years of arduous labor and for heavy expenses incurred in the service of the republic.

In advance of the negotiation of the loan, steps were taken to build two small steamers, 125 feet in length, and to transport them in pieces to Bolivian waters above the falls of the Madeira and Mamoré. On his return to England, Colonel Church found new and unexpected obstacles confronting him on every side. France had declared war against Prussia, and the financial world was in such a feverish state of excitement and apprehension that it seemed useless to attempt a negotiation of the proposed loan at that time. It was found that many years before, England had practically discontinued diplomatic relations with Bolivia in consequence of the bad treatment received by a British charge d'affaires in 1853. In December, 1870, Mr. Brassey died, and Mr. Wheelwright was not disposed to proceed with the enterprise alone. Colonel Church had relied upon the vast wealth and high reputation of these men as the most satisfactory pledges that could be offered for the faithful and energetic execution of his project.

In his effort to secure other testimony than his own to the great natural resources of Bolivia and the commerce likely to be developed by the construction of the Madeira and Mamoré Railway, Colonel Church found little or nothing, recently published to substantiate his own statements, and he tells us that "Bolivia was far less known to Europe in 1870 than any country on the globe having a regular form of government." Several previous attempts by Bolivia to float loans in Europe had proved unsuccessful and that country's credit had been seriously impaired, years before, by a scheme that attempted to mortgage to Messrs. Peto, Betts, and Co., all Bolivia's resources—as Colonel Church expressed it, "the whole of Bolivia, animal, mineral, and vegetable—financial and almost spiritual—exclusive rights of internal communication and river navigation—Bolivia inside out—Bolivia present and future—Bolivia down to the very base of the Andes."

Such difficulties might well discourage and dishearten any ordinary man, but the same industry, courage, and self-reliance that characterized his Pilgrim ancestors was exhibited in the conduct of Colonel Church, their worthy descendant.

The same unceasing energy, indomitable will, and inflexible determination he had manifested as a boy at home, on the pampas of Buenos Aires, through the dark days of the Civil War, and on the mountains and plains of Mexico guided his conduct then and in every succeeding emergency. On March 1, 1871, the Madeira and Mamoré Railway Company, Ltd., was incorporated with Colonel Church as chairman of the board of directors and Mr. George Hopkins, CE, as chief engineer. This company purchased the Brazilian concession from Colonel Church for twenty thousand pounds cash. Through representations made by Colonel Church to the government of Bolivia, diplomatic relations with England were resumed, and, with private funds, he bought the notorious claim of Peto, Betts, and Co. for eight thousand pounds. Through newspaper and magazine articles, he made known the great natural wealth of Bolivia and her need for an adequate commercial outlet. He submitted his financial proposition to the great bankers of Paris, Brussels, Amsterdam, and finally, to Erlanger and Co., of London and Paris.

Later came the news of horrible barbarities committed by the La Paz Indians in Bolivia—for example, that they had buried Minister Muñoz—the government was powerless to suppress the insurgents under Morales, and President Melgarejo had fled. This news was only partly true, but Messrs. Erlanger and Co. would consent to further consider the proposed loan only on condition that Colonel Church at once proceed to Bolivia and secure a confirmation of the loan contract by the new government of President Morales.

This firm had been convinced of the merits of the enterprise and had agreed to issue the proposed loan, when the news of the revolution in Bolivia reached England, upset all arrangements previously made, and compelled Colonel Church's immediate return to that country.

Before his departure, contracts with Messrs. Erlanger and Co. and the Public Works Construction Company were drawn up and signed on May 18, 1871. These contracts were as follows:

>   (1) For the issue of a loan by Erlanger and Co., signed by Colonel Church as special agent of Bolivia;
>
>   (2) For the disposal of the proceed of the loan;
>
>   (3) A deed of security from the National Bolivian Navigation Company, appointing trustees;
>
>   (4) A contract for the construction of the Madeira and Mamoré Railway, between the navigation and railway companies and the Public Works Construction Company.

Messrs. Erlanger and Co. made it a condition of issuing the loan that the Public Works Construction Company should build the railway. The high financial standing of the company and the eminently respectable character of its board of directors secured the ready acceptance of this condition by Colonel Church and the interests he represented. The Public Works Construction Company decided to have the railway route examined by two of its own engineers, Mr. C. F. de Kierzkowski and Mr. Leathom Earle Ross.

Colonel Church, accompanied by Mr. Ross, sailed for New York two days after the execution of the contracts before referred to and, after a short stay in that city, on June 15, 1871, sailed for Bolivia via Panama and Peru.

They reached Sucre on July 30, 1871, and by August 28, the loan contract with Erlanger and Co. had been confirmed by the Bolivian Congress and approved by President Morales. Three days later, the formal documents attesting such confirmation were handed to Colonel Church and, with Mr. Ross and three others, he at once set out for Cochabamba to study that place in connection with the navigation enterprise.

September 10 found the party en route for Santa Cruz de la Sierra, which it reached on the nineteenth of the same month.

Colonel Church found all the towns along his route enthusiastic for the opening of river navigation. Leaving Santa Cruz on September 27, the party proceeded to the little port of Cuatro Ojos on the Piray, a branch of the Mamoré River. There, on September 30, the original party, increased by thirty-seven Indian boatmen and eleven other persons, embarked in canoes. By October 6, 1871, they had reached Trinidad on the Mamoré.

Previous to Colonel Church's departure from New York, the navigation company had purchased a small schooner, the *Silver Spray*, upon which they intended to load the pieces of the steamer *Mamoré* and convey them, with a force of workmen and mechanics, to San Antonio. There it was proposed to await the arrival of an expedition to be organized in Bolivia for transporting the pieces above the falls and rapids, where the mechanics were to put them together and launch the little steamer on the waters of the Mamoré River. Pending the loan negotiations in London, a forty-foot iron steam launch, the

Explorador, had been constructed for use on Bolivian rivers and conveyed to South America.

At Trinidad, therefore, the final arrangements were made by Colonel Church for the more dangerous part of his voyage, through the obstructed portion of the river, and for the return trip with the *Explorador* and *Mamoré*. After leaving Trinidad on October 11, with four boats and a party of eighty-three men, he had a short stay at Exaltacion and then successfully passed the falls and rapids and halted his party on the west bank of the Madeira opposite San Antonio. There, he met an American, Silas S. Totten, who had gone to Bolivia in the employ of the navigation company and returned with twenty-eight Indian workmen to San Antonio, where he had been awaiting further orders for some time. On November 1, the site of the railway terminus was selected and, in the presence of the entire party, consisting mainly of half-clad Indian boatmen, in the heart of a vast tropical wilderness, fifteen hundred miles from civilization, where the interior country was absolutely unexplored and, except by primitive savages, uninhabited, Colonel Church went through the rude ceremony of turning the first sod for a railway, that, with the faith of a Columbus, he firmly believed and fondly hoped would open to immigration and to the commerce of the world a country unsurpassed in latent wealth by any unoccupied territory of equal extent on the face of the globe. No more inspiring subject for poet's pen or painter's brush could be suggested than the scene here presented by this soldier and citizen of the great Republic of the North, standing amid the wild surroundings at the head of navigation on the Madeira and pointing out to the sister Republic of the South, for centuries the victim of foreign oppression and internal strife, the pathway of future peaceful progress and commercial development.

Notwithstanding all the misfortunes that subsequently attended the enterprise, it is safe to predict that this event will yet be regarded as marking the beginning of a new epoch in South American history.

It was arranged that Mr. Totten should remain at San Antonio to clear the ground, erect houses, and in a general way, represent Colonel Church's two companies.

The men who were to take the *Explorador* to Bolivia were also placed under his charge. Just as Colonel Church was about to embark, the *Explorador* came steaming up the river and brought the news that the *Silver Spray* with her freight was at Serpa on the Amazon near the mouth of the Madeira, but that, "owing to the death of her captain and some of her officers, could not continue her voyage to San Antonio."

The *Explorador* carried Colonel Church and some of his party to Serpa, where they arrived on November 10. There, Señor Velarde was detached with instructions to return to San Antonio on the *Explorador* and, with the force left there, take her above the obstructions of the Madeira and Mamoré, a most difficult undertaking. While making the necessary portages, Señor Velarde was instructed to make a good road around each fall, in anticipation of another expedition, which he was expected to organize in Bolivia for the purpose of conveying the pieces of the *Mamoré* over the same route. Colonel Church and Mr. Ross boarded an Amazon steamer at Serpa and reached London on December 14, 1871.

Mr. Ross estimated the cost of the railway at £437,989.

A final contract was entered into with the Public Works Construction Company to build the railway for six hundred thousand pounds, upon which the company demanded and received an advanced payment of fifty thousand pounds. The Bolivian loan was placed upon the market by Erlanger and Co., January 20, 1872, and bonds to the nominal value of £1,700,000 were sold at 68 percent. The total amount realized, after paying all expenses and a five percent commission to Erlanger and Co., was £1,040,286. This sum was appropriated as follows:

Set aside for sinking fund and interest	£187,000
Placed to the credit of Bolivia	145,058
Deposited to the order of railway trustees	600,000
Paid to the National Bolivian Nav. Co.	108,228
Total	£1,040,286

The navigation company had agreed to convey all the men and material to San Antonio, and for that purpose, a powerful iron side-wheeled steamer and ten iron schooner-rigged barges of 250 tons each were contracted. The steamer was intended for towing on the Amazon and Madeira, the barges for transportation in pieces to Bolivian waters. A dozen or more wooden launches were ordered to be constructed on the Mamoré River to aid in bringing labor and provisions from Bolivia to the scene of operations. A steamship was chartered at Liverpool, and a river steamer was sent to the Amazon to run between Pará and San Antonio. The contractors organized and dispatched to San Antonio a corps of twenty-five engineers under Leathom Earle Ross, CE, and the railway company sent to the same point Mr. Edward D. Mathews, as a resident engineer, with some assistants. The entire force of engineers and contractors reached San Antonio July 6, 1872. Colonel Church again returned to Bolivia, reaching La Paz July 26, 1872, and there made full report to the government of all that had been done. While at La Paz, he learned that Señor Velarde had succeeded in placing the *Explorador* at El Cerrito, near Exaltacion on the Mamoré. This was no trifling feat to accomplish and involved dragging the iron launch for miles on land over rocks and through dense tropical forests in the rainy season and at a time when fevers were most prevalent and virulent. The machinery of the launch had been very much impaired by the strain to which it had been subjected, but the hull was still in serviceable condition.

After a stay of four months at La Paz, Colonel Church set out for Rio de Janeiro via the Straits of Magellan in order to settle a difficulty that had arisen in regard to carrying freights on the Amazon and Madeira under the British flag and to secure a hearty cooperation between the Brazilian government and the contractors.

Colonel Church remained at Rio de Janeiro three months, during which the city was visited by one of the worst yellow fever epidemics that had ever occurred on that coast.

On his return to England, he found that, a year from the date of their contract, the contractors had accomplished almost nothing. A quantity of lithograph plans and three and a half miles of survey on the line of the road was practically the whole work accomplished. Not a foot of track had been laid, no grading had been done, and not a sleeper had been cut or delivered.

On July 9, 1873, one year and three days after its engineers reached San Antonio, the Public Works Construction Company, finding the difficulties encountered greater than had been anticipated and determining that it would be impossible to complete the road within the time or for the price agreed upon, definitely repudiated the contract and filed a bill in Chancery petitioning the court to declare the contract null and void, asking for reimbursement for expenditures made and seeking to restrain the trustees from making further payments out of funds in their charge, until the contractors' claims could be adjudicated. The construction company claimed that the whole enterprise and particularly the length of the road had been misrepresented to them, "that the country was a charnel house, their men dying off like flies, that the road ran through an inhospitable wilderness of swamp and porphyry ridges alternating, and that, with the command of all the capital in the world and half its population, it would be impossible to build the road."

The effect of such representations from a seemingly authoritative source was disastrous in the extreme. The Bolivian government was under obligation to pay six percent interest on the bonds issued and had pledged the faith, credit, and "revenues and possessions" of the republic, not only for the continued payment of the interest, but also for the final extinguishment of the loan, yet nothing had been done and, accepting as true the statements of the construction company, no advantage to Bolivia was ever likely to result from the expenditure.

The Bolivian government had relied upon the increased revenue resulting from the opening of navigation on her rivers and the construction of the railway to pay interest on the Church loan and eventually that loan itself. The prospective earnings of the navigation company constituted one of the most important securities pledged to the bondholders when the loan was placed upon the market. If the construction of the railway was impracticable, as alleged, the bondholders had little security beyond the faith, and credit of the Bolivian Republic, which, in the absence of any present or prospective return, found the financial obligations it had assumed extremely burdensome.

The bonds, in consequence, depreciated from sixty-eight to eighteen cents on the dollar. The original bondholders were not in an enviable position, and, to complicate matters further still, a group of shrewd London brokers and speculators stepped in and bought up about two-thirds of the entire bond issue at a nominal price, expecting to realize an enormous profit in the event that the construction of the railway was declared by court impracticable and the trust fund in the Bank of England, then amounting to seven hundred thousand pounds, ordered to be divided among them. In order to give a death blow to the ruinous allegations of the Public Works Construction Company, without waiting for the results of tedious litigation, the railway and navigation companies, on September 17, 1873, entered into a new contract for the construction of the road with Messrs. Dorsey and Caldwell, American contractors of skill and experience in such work, who agreed to construct the first and worst ten miles of railway before asking any remuneration whatever.

Litigation made impossible any prompt prosecution of the work by this firm, and finally, tired of the delay, these gentlemen, on August 25, 1875, transferred their interest in the contract to Reed Bros. and Co. of London with the consent of all parties concerned. One would naturally suppose that in the contest between Colonel Church and the Public Works Construction Company all the influence of Bolivia, which was so vitally interested in the success of the enterprise, would be exerted on the side of the former, but, unfortunately, at this time, Señor D. Adolfo Balivan, who had been acting as the London agent of Bolivia in an unsuccessful attempt to negotiate a new loan, was elected president. Realizing both the necessity and the difficulty of replenishing the exhausted Bolivian treasury, he conceived the brilliant idea of canceling the Church enterprise, buying up the bond issue at the low price it had then reached, and confiscating the proceeds of the loan negotiated by Colonel Church. Having thus wiped out all Bolivia's obligations under the loan contract and, having incidentally, given a death blow to Bolivian credit, he was possessed of the insane idea that he could then float on the London market a new loan for fifteen million dollars that would be sufficient to meet all outstanding obligations of the government. The people and the National Assembly of Bolivia were enthusiastically in favor of the railway and navigation projects. This fact is abundantly proved by the action of the municipal council of Cochabamba, which passed resolutions heartily endorsing the Church enterprise and calling upon the government "to facilitate Colonel Church, a friend of Bolivia, in his action against the Public Works Construction Company," by the course of the people of Santa Cruz, in sending Colonel Church a vote of confidence, and by the proceedings of the National Assembly.

On June 11, 1873, influenced by President Ballivan's representations, the Bolivian National Assembly passed a secret decree authorizing the president to *guard* the funds in England resulting from the loan of May 1871, by means of agents in whom he had confidence and to examine the navigation enterprise with the object of securing all the amounts invested, in case the contractors did not faithfully comply with their obligations.

The deputies failed to pass the new loan bill contemplated by the president so the action taken did not accomplish the purpose he had in view. The president then convoked the assembly in extraordinary session on October 8, 1873, and the administration met with a second defeat. Instead of approving the loan scheme, the assembly passed a decree directing the executive to aid, by all means in his power, the railway and navigation enterprises and authorizing him, if necessary, to extend the time for completing the projected improvements. Unfortunately, without intended hostility to Colonel Church's projects, the assembly invested too much discretionary power in the president and the decree required that in case the navigation company failed to give such security for the completion of the work *as the executive deemed sufficient*, the government should intervene and endeavor to recall the moneys proceeding from the loan negotiated by Colonel Church. In advance of this action by the assembly, President Ballivan had dispatched agents to London authorized "to take possession of and secure the existing funds of the Church loan and those invested in the purchase of materials in any part of the world, wheresoever found, and to deposit them in the Bank of England as a guarantee of said loan ... because later it may be opportune to order the regulation and amortization of these bonds on a convenient basis."

As a result, Colonel Church found all the powerful interests, which would naturally be allied with his own, arrayed against him and, strange to say, earnestly endeavoring to sustain and maintain the contentions of the Public Works Construction Company, which had been the sole cause of delay in the execution of his projects. He made a futile endeavor to arrive at an early decision of the questions at issue by arbitration, but his proposals did not prove acceptable to his opponents, and, after some preliminary skirmishing, he entered the legal contest with Bolivia, the bondholders, and the defaulting contractors. On August 7, 1873, the companies Colonel Church represented brought a countersuit for damages, resulting from breach of contract, against the construction company. On the twentieth of the same month, they filed a friendly bill in Chancery against the trustees of the fund in the Bank of England and made the bondholders parties to the suit, the purpose being to free the trust fund and make possible the continuance of work on the railway. On January 15, 1874, the bondholders filed a bill against the railway and navigation companies, the custodians of the trust fund, the Republic of Bolivia, and Colonel Church, seeking to restrain the trustees from making further payments out of the trust fund and to have that fund, with accumulated interest, declared by the court to be the property of the bondholders, in case the construction of the railway was pronounced impracticable. On May 4, 1874, the Republic of Bolivia filed a bill against the railway and navigation companies, the trustees, and Colonel

Church, praying that the trust fund be declared the property of the Bolivian government.

After the repudiation of its contract, the Public Works Construction Company gave the only evidence of energy it had at any time manifested in connection with the enterprise by having its engineers run a crooked line between the terminal points of the railway. Their object was to obtain legal evidence that the distance was greater than had been supposed, and in this, apparently, they were successful.

It would be neither entertaining nor profitable to trace the litigation that followed through its entire course and in all its details. During the eight years that elapsed between the inception of the enterprise in 1868 and the decisions handed down by the court in November 1876, Bolivia had five presidents and, of these, the three last were unmistakably hostile to the construction of the railway. The fund of seven hundred thousand pounds lying idle in the Bank of England was viewed with longing eyes by these rulers and, unable to find means at home to gratify their avarice and cupidity, they did not hesitate to even exceed the powers given them by law in their efforts to divert the money to other purposes than those for which it had been raised and to the accomplishment of which it had been solemnly pledged.

That the legal position assumed by the representatives of Bolivia was not regarded by themselves with unshaken confidence is made evident by the fact that President Frias offered eighty thousand pounds and President Daza two hundred thousand pounds to Colonel Church on condition that he would abandon the enterprise and aid in their nefarious schemes.

Colonel Church soon realized that procrastination and delay were the policy of those allied against him, that the allegations of the Public Works Construction Company constituted the real basis of the entire litigation, and that, regardless of any judicial decision on the issues, the law's delay would work irreparable injury to the interests he represented. He, therefore, felt justified in making a compromise on October 13, 1874, with the Public Works Construction Company, by the terms of which that company withdrew all charges of misrepresentation and accepted forty-five thousand pounds and costs in consideration of expenditures claimed to have been made. In the suit against the trustees to free the trust fund, the master of rolls, on December 11, 1874, handed down a decree stating that "The Plaintiffs are entitled to have the United States bonds ... deposited in the Bank of England ... and the interest accrued and accruing thereon applied from time to time in payment for the works of the railway ... as such works proceed and the trustees ... are bound to apply the same accordingly."

On June 27, 1876, the Bolivian minister at London declared the Bolivian concession null and void and announced the withdrawal of his government from any further recognition of the railway and navigation enterprises. This act was clearly a usurpation of authority that could be lawfully exercised only by the Bolivian National Assembly.

On November 22, 1876, the bondholders' suit was dismissed and the costs placed on the plaintiff, the master of rolls declaring, "There is no equity in this suit." On the same day, the same judge dismissed the bill filed by the Bolivian government at the cost of the plaintiff.

The law allowed one year in which to appeal these decisions, and in one case, a sham appeal was entered within a week of the expiration of that period but was afterward withdrawn.

The compromise and three decisions by the court seemed to remove all obstacles to the prosecution of the work, and on November 27, 1876, Reed Bros. and Co. was notified to proceed with the execution of their contract. A long correspondence resulted. The firm made a demand for an advance payment, a remodeling of their contract, and a stipulation from the trustees that they would apply the trust fund as ordered by the court. The firm would neither abandon its contract nor proceed with the work. Finally, on January 18, 1877, under legal advice, Colonel Church declared the contract null and void. Then Reed Bros. and Co. began an action for damages. The writ was issued February 6, 1877, and later Vice Chancellor Bacon decided the notice abrogating the contract premature and authorized an inquiry to determine the amount of damages. A compromise was then reached fixing the sum due the contractors at twenty-five thousand pounds.

The abandonment of the enterprise by the two English firms and the loss of time caused by litigation had been ruinous both to the railway and to the navigation company, and had involved both in heavy expense. The many boats contracted for had been sold at a great sacrifice. The *Explorador* from rust and want of care had become worthless at El Cerrito. Houses and shops erected at the same point, as well as others at San Antonio, had fallen into decay and ruin. Roads cut around the falls and in other places had become so overgrown that it was difficult to find them. The *Silver Spray* had sunk at San Antonio with all her cargo, including the pieces of the *Mamoré* on board. Besides legal expenses incurred, the railway and navigation companies had been under heavy expense to maintain necessary agencies at various points. The bondholders had accomplished nothing except to depreciate the market value of their bonds and impair the only good security that existed for the ultimate extinguishment of the loan. The Bolivian government, at considerable expense and trouble, had seriously injured Bolivian credit. Until June 1875, the bondholders had been paid interest out of the money subscribed by them, but after that date, interest payments ceased, and the debt, which Bolivia was obliged to pay, was piling up at the rate of £136,000 annually, while that country was fighting to prevent the accomplishment of the purposes for which the loan had been authorized. In short, the only parties involved in the litigation who in any way benefited from the result were the defaulting contractors, who should have been made to pay heavy damages instead of receiving a reward.

During the early stages of the litigation, however, it became evident that the Brazilian government was ready to aid the enterprise financially the moment the legal obstructions were removed. As early as October 1874, the legislature of Brazil was considering the guarantee of additional funds for the construction of the railway, and on March 8, 1876, Colonel Church received authentic information from Rio de Janeiro to the effect that a bill had passed the legislature authorizing a guarantee of seven percent interest on four hundred thousand pounds of additional capital, to be expended in the construction of the railway after the trust fund in England was exhausted. The emperor had attached his signature to the bill, but the authority to give such a guarantee was not to be utilized until the legal obstacles to the prosecution of the enterprise had been removed. Subsequently, this guarantee was promulgated by the imperial decree of November 24, 1877, and published in the *Official Journal* of the Brazilian government. On October 18 of the same year, the emperor of Brazil issued a decree extending the time for the construction of the railway, thus attesting his faith, as emphatically as possible, in the "impracticable" enterprise in which Colonel Church was engaged.

- Dom Pedro II, having become emperor of Brazil in July of 1841 at the age of fifteen, had a deep appreciation for science and technology. Along with his international connections, he brought domestic peace, prosperity, and progress to Brazil during the near half-century of his reign.
- During his reign, "Union and Prosperity" brought the first paved-stone road linking Rio de Janeiro, Petropolis, and Juiz de Fora; the first steam-engine railroad started running between Santos and Sao Paulo in 1868; the Brazil-Europe submarine cable was installed; and the postal stamp was instituted.
- In 1852, he granted permission to the US Navy to survey the Amazon River Complex, from the headwaters to the Atlantic Ocean, in preparation for international trade.
- Dom Pedro II visited the United States in 1876 to join President Ulysses S. Grant (1822–1885) in opening the

Philadelphia Centennial Exposition, the largest world's fair up to that time.
- He had composed a complete musical for the occasion and then directed the National Orchestra in tribute to the fledgling United States of the Americas.
- Dom Pedro II was the first foreign head of state to visit the United States and formally addressed the US Congress.
- At the Centennial Exposition, he and President Grant jointly threw the switch on the Corliss Engine, thereby providing the power for the entire exposition. With a congressman from Western Pennsylvania, he toured the railroad transportation system, the famous Horseshoe Curve, the railroad tunnels, and the only "inland-port" of that day.
- At the exposition, he met and talked with Alexander Graham Bell, who demonstrated his latest invention, the telephone. Dom Pedro recited Shakespeare's classic line from *Hamlet*, "To be or not to be," into it and exclaimed, "This thing speaks!"
- He thereupon invested ten thousand dollars and became a founding investor in the Bell Telephone Company. Upon returning to Brazil, he installed the country's first telephone in his family's summer retreat in Petropolis, forty miles from Rio de Janeiro.
- He was interested in expanding his country's primary agriculture commodities and in making Brazil a major cotton producer. After the defeat of the Confederate States in the US Civil War, he invited successful Confederate cotton planters to settle in Brazil. Between

1867 and 1871, when slavery was still legal in Brazil, at least three thousand Confederate families passed through the port of Rio de Janeiro. About 80 percent of the families later returned to the United States, but one successful American settlement in Brazil, Americana, founded by Colonel William Hutchison Norris (1800–1893) of Mobile, Alabama—still exists. Located seventy-five miles from Sao Paulo, Americana in 2003 has a population of approximately 250,000 persons. Conversations among the descendants of the *"confederados,"* who compose about 10 percent of the population, are often in Southern-accented English. Families with names such as Jones, MacKnight, and Whitaker come together for the Fourth of July and other holidays to have a Southern-style barbeque. Rosalyn Carter, wife of President Jimmy Carter, has relatives buried in the Confederate Cemetery in Americana. In 1992, President-elect Bill Clinton wrote a letter to the *confederados* of Americana in which he recalled that Arkansas was one of the thirteen states that had sent settlers to Brazil.

- Liberal in outlook, Pedro II, a strong admirer of Abraham Lincoln, took steps to end slavery, and the final abolition edict was signed in his absence because of his declining health by his daughter Princess Isabel on May 31, 1888.
- Also at this time, he was in the process of building the first opera house, the Theatre do Pas, in the Amazon region in the city of Belem in Para State, which was completed in 1889. He willingly gave up his reign as

emperor as part of his belief in Brazil's evolution, and in 1889, the republic, the United States of Brazil, was born. In 1989, the State of Para built the FEMAZON State Fair, in honor of his dedication to Brazil's future.

*

IV. The Organization of the American Expedition

Make the charred logs burn brighter;
I will show you, by their blaze,
The half forgotten record
Of bygone things and days.
—Adelaide Procter

The decisions handed down by the master of rolls on November 22, 1876, made it necessary for Colonel Church to look for new contractors so that at the expiration of the year allowed for appeal, he would be prepared for a prompt and vigorous prosecution of the work. It is not surprising, after his discouraging experience with English contractors, who had accomplished practically nothing except to involve them in heavy expense and years of litigation, that he then sought in his native land for men of greater nerve and determination. He was, of course, familiar with the work accomplished by American engineers at home, in Mexico, in Peru, and on the Isthmus of Panama, so it was only natural for him to suppose that equal success would attend the employment of American engineers and American methods on the upper Madeira.

About this time, Mr. Dillwyn Parrish, a London contractor, who was a director in the railway company, happened to be in the United States. Colonel Church cabled the result of the litigation to him. Mr. Parrish explained the situation of the two companies to Colonel John Jameson of Philadelphia and to Mr. Hugh Hamilton, a commission merchant of the same place, and stated that, under the decisions rendered, it would soon be possible to enter into a new contract for the construction of the Madeira and Mamoré Railway.

These gentlemen, in turn, communicated the facts, thus obtained, to P. and T. Collins, railway contractors of Philadelphia, and they at once became deeply interested in the project. Messrs. Philips and Thomas Collins, the members of this firm, were well and favorably known to every railroad man throughout the State of Pennsylvania. Thomas Collins had been an employee on the old Pennsylvania Canal, and one or both members of the firm had been identified with almost every important piece of railway construction in Pennsylvania that had been undertaken during the previous twenty-eight years.

Their first railway contract was for the building of the old Portage Railroad, which was opened for traffic in 1849. Later, they constructed the Philadelphia and Erie Railroad, part of the main line of the Pennsylvania Railroad, including the long tunnel at Gallitzin, the branches to Ebensburg and Tyrone, and the Clearfield and Indiana branches. After other contractors had sunk large sums of money in it and failed, P.

and T. Collins took the contract to complete the celebrated Sam Patch Tunnel near Cumberland, Maryland, and were successful. Afterward, they executed large contracts for the Philadelphia and Reading Railroad and the Lehigh Valley and Jersey Central. They also built a part of the Western Pennsylvania Railroad, had a large contract on the Southern Pennsylvania, and constructed most of the Beach Creek, as well as a part of the Lewisburg and Tyrone and many private railroads.

Mr. Franklin B. Gowen was then in London in response to a cable inquiry of the Madeira and Mamoré Railway Company to pay for the work. At the same time, Mr. Gowen perceived that, by securing the contract for P. and T. Collins, he might greatly increase the business of the Philadelphia and Reading Coal and Iron Company, of which he was president. On August 28, 1877, the Collins brothers made an agreement with Mr. Hugh Hamilton by which, "in consideration of such information as had already been given and assistance which may be given and services to be rendered" in personally negotiating the contract with the Madeira and Mamoré Railway Company, Mr. Hamilton was to receive one-tenth of the net profits under the contract. There is reason to believe that an equal interest, for similar considerations, was given to Colonel Jameson at about the same time. Neither of these gentlemen invested any money in the enterprise, though both contributed personal services, and from his signature to several papers, it is evident that Mr. Hamilton held a power of attorney to represent P. and T. Collins in London during the negotiation of the contract.

The result was that, in a very short time and largely

through the influence and representations of Mr. Gowen, P. and T. Collins secured the contract to build and equip the railway. This contract, dated October 25, 1877, provided, among other things, that work on the ground at San Antonio should be commenced within four months from that date and that the road should be completed within three years. This elaborate document was in some respects peculiar. While the road was to be of one-meter gauge and was only required to be capable of carrying trains of two hundred tons at a speed of twenty miles an hour, the minimum radius of curvature was fixed at 1,320 feet and the maximum gradient at one in one hundred. There was, however, a provision that, under exceptional circumstances and by special permission, curves of six-hundred-feet radius and gradients of one in fifty might be used. The contractors were required to deposit United States government bonds to the amount of two hundred thousand dollars with McCalmont Bros. and Co. of London, as security for the faithful performance of their contract obligations. While it was specified that the contractors were to be paid £5,900 per mile of main line for 180 miles and £5,200 for each mile in excess of that distance, the qualifications in the contract were such that the contractors might receive considerably more or less than this mileage price, according to the amount of grading, ballasting, masonry, and bridgework required, these items being virtually paid for by quantity at specified rates. The contractors were to spend fifty thousand pounds in "conveniences" without charge and, if required to spend more, were to receive actual cost and fifteen percent additional, as compensation for the excess. The main line was to have 3 1/2 miles of sidings without additional charge, except for grading and masonry, which were to be paid for as extra work at fixed prices, according to quantity. Ballast in station grounds and sidings to the amount of 8,480 cubic yards were to be supplied without extra charge, but any excess was to be paid for at a stated price.

Contracts, as a rule, are far from interesting, but it is absolutely necessary, in order to appreciate properly the difficult position in which the contractors were subsequently placed, to refer to the provisions of no less than four other contracts made by P. and T. Collins about the same time.

The compromise that had been made with the Public Works Construction Company required the payment of forty-five thousand pounds, but as this settlement was a private arrangement made to terminate ruinous litigation, there was no possible way in which that sum could be taken from the trust fund directly. Accordingly, on the same date as the contract for the construction of the railway, the new contractors, by a special agreement, assumed all rights and obligations of the railway and navigation companies under the compromise made with the Public Works Construction Company. This gave P. and T. Collins a right to all the material left at San Antonio and to some maps, plans, and other papers of little or no value. However, they were under obligations to permit a deduction of 7.5 percent from their own earnings for the benefit of the Public Works Construction Company, until the entire sum of forty-five thousand pounds was paid.

By another agreement, made October 25, 1877, with the National Bolivian Navigation Company, P. and T. Collins bound themselves to purchase from the said company, at their full nominal value of twenty pounds per share, 2,500 shares of stock in the railway company in fixed installments at certain specified dates extending over a period of three and a quarter years. The object of this agreement was probably to furnish the navigation company with funds to pay salaries and other expenses.

It is probable that P. and T. Collins were not prepared at once to deposit the two-hundred-thousand-dollar security required by the construction contract. However that may be, on October 30, 1877, they entered into an agreement with McCalmont Bros. and Co. under the terms of which these bankers were to furnish the required security for nine months and at the end of that time it was to be repaid by the contractors with interest. As compensation, the bankers were to receive 1 percent of the gross receipts of the contractors, not only during the nine months, but throughout the whole duration of the contract. Furthermore, as security, McCalmont Bros. and Co. were appointed irrevocably attorney-in-fact for P. and T. Collins with full power to collect all sums due them under the railway contract and, after deducting the 1 percent previously mentioned, to retain one-half the remainder until the entire amount of two hundred thousand dollars and interest was repaid. Finally, the contractors, on October 30, 1877, made an agreement with the Philadelphia and Reading Coal and Iron Company, which provided that, in consideration of assistance given by the said company to P. and T. Collins in obtaining their contract and making arrangement for the required security through McCalmont Bros. and Co., the contracting firm was to purchase from the Reading Coal and Iron Company all the ironwork and rolling stock of every character and description required in the construction and equipment of the railway at actual cost and 10 percent additional. Further, the contractors agreed to pay the company the sum of $150,000, and Franklin B. Gowen was appointed irrevocably attorney-in-fact for the contractors, with power to collect all money due them from McCalmont Bros. and Co. From such collections, Mr. Gowen was authorized to deduct 3 percent until all payments due to the Reading Coal and Iron Company had been made.

For reasons unknown to the writer, several other persons, besides those already mentioned, were admitted by the Collins brothers to partnership in the prospective profits of the enterprise. The respective interests were as follows: P. and T. Collins, two-fifths; Jameson and Hamilton, one-fifth; Josephs and Barger, one-tenth; Frank McLaughlin and Bro., one-fifth; and James Scott of Mackie, Scott, and Co., Ltd., one-tenth. According to subsequent testimony in the English courts, Josephs and Barger invested about sixty-five thousand dollars in the undertaking, but, so far as is known, they were the only partners of P. and T. Collins who acquired their interest by pecuniary investment.

With reference to the construction contract, the *London Railway News* made the following comment:

It is to be regretted that a work, in which so many English interests are involved, has been taken out of English hands, but the bondholders, who have been waging a long contest with Colonel Church, the agent of the Bolivian Government, have themselves to blame for the action he has at last taken in seeking in America the support denied him here in carrying out the scheme for which funds were duly provided and have been so long lying idle in the Bank of England.

Pennsylvania's Amazon Princess Railroad

Under the circumstances it is a satisfactory feature that the work has been undertaken by a well-known and enterprising firm of contractors, in whose hands there seems every probability of a speedy completion of a work of vast importance to the development of South American trade and which promises to prove a remunerative investment.

As an examination of the old files of Philadelphia newspapers would seem to indicate, the first news of the Collins brothers having secured this large contract reached this city on or about November 7, 1877, and from that time on, the Brazilian enterprise was the subject of numerous newspaper articles, many of which displayed the most laughable ignorance of South American geography and by too highly coloring some facts and omitting others, rather than by any positive misstatement, conveyed the impression to the reading public that the building of the railroad was to be the occasion of a mere picnic excursion that would enrich every one of the participants at the cost of little labor and exertion on their own part. Frequently, a newspaper would flatly contradict on one day the statements it had made only a few days before. "The six-million-dollar contract" was on every tongue. The effects of the financial panic in 1873 had not yet ceased to be felt, and in every sizable city in the East, large numbers of unemployed workmen were to be found. The practical suspension of extensive railroad building, which in 1873 had reached previously unheard of proportions and caused an abundance of unemployed and experienced civil engineers, contractors, and mechanics, who now hastened to assure the Messrs. Collins that they were open to an engagement. Fathers returned from business at night to tell their sons of "the El Dorado of the South" they had just discovered in the columns of an afternoon newspaper. The whole city was in a ferment. It was more difficult to secure an interview with the Messrs. Collins than with the president of the United States. The *Times* of November 8 said, "They [Messrs. P. and T. Collins] have been in receipt of eighty thousand or less applications for positions on the work." Laborers by the hundred congregated in front of the contractors' offices in Walnut Place and remained there from early in the morning until late at night. Contractors from all over the country rushed to this city in an endeavor to secure subcontracts on the work.

On December 6, one of the city newspapers published a long interview with Colonel Church, which concluded with these words:

> It is to remedy this (the commercial isolation of Bolivia) and to open up to the world a land as fair as "the Garden of the Lord" that two Philadelphians are to overcome the rapids of the Madeira. "I am not visionary, but know whereof I speak," said Colonel Church, "that this once accomplished, the wealth of Australia and California will sink into insignificance beside the auriferous yield of the mountains and streams of Bolivia and the teeming products of her fertile plains and valleys.

"Tom" Collins, as his intimate friends familiarly called him, was the active spirit and executive officer of the firm.

Whatever personal peculiarities he may have been charged with, not even his bitterest enemy ever accused him of frittering away valuable time on newspaper interviews and books of travel, when important work had to be done and done promptly to comply with contract requirements.

He alone, of all Philadelphia, was equally indifferent to glowing descriptions of "the Garden of the Lord" and to the somber shadows thrown over Colonel Church's enthusiasm by some pages in the reports of Gibbon and Keller. If he took the trouble to compare the statements, he was probably convinced, as were many others, that one side or the other was drawing upon their imaginations for their facts and that the imaginative individuals were just as likely to be Gibbon and Keller as Colonel Church and his associates. His whole life had been one of strenuous action. He was a big man physically, with a heart proportionate to his frame, and, like many big men, seemed to have a feeling that mere physical energy would overcome every obstacle.

It was an unfortunate circumstance that, during the time of preparation, the contractors had in their employ no men with railroad experience under the peculiar conditions to be encountered, but, in justice to them, it must be recollected that such men were not obtainable, unless from the Public Works Construction Company of London, which had already most thoroughly demonstrated its incapacity for such work.

The energy displayed in raising, equipping, and transporting the first installment of men and material was something phenomenal. In exactly fifty-four days from the time Philadelphia first heard of the contract, the steamer

William Lawrence Adams

Mercedita, with Captain Jackaway on board and steam up, was impatiently waiting at the Willow Street wharf for the legal holiday to pass before starting on her long voyage to San Antonio.

V. The Voyage of the *Mercedita*

White curl the waves,
and the vexed ocean roars.
—Pope

A correspondent of the New York *Herald*, writing from Philadelphia on January 2, 1878, said:

A national interest centers in the voyage of this ship, for the reason that it is the first time in the history of this country that an expedition has been sent from the United States, equipped with American money, materials, and brains, for the execution of a great public work in a foreign country. ... The party of engineers, fifty-four in number, is said to be the ablest body of men of this profession ever united in a similar expedition.[8]

The *Mercedita* was a screw steamer of 856 tons burden. She had been launched in 1852 and during the Civil War, as

---

[8] "When finished it will be the only railroad outside of the United States constructed from end to end by Americans and ironed and stocked with American rails and rolling stock," from an interview with Colonel George Earl Church in the *Philadelphia Times*, Dec. 6, 1877.

a gunboat, had aided in the blockade of Confederate ports. Subsequently, she had been converted into a merchant vessel and engaged in trade between New York City and St. John, New Brunswick.

Her master, William Jackaway, was a veritable seadog, whose life had been spent in the whaling trade and, as appeared later, was not any too familiar with the route the *Mercedita* was about to follow.

Colonel John Jameson was general manager of the entire expedition, in the absence of Mr. Thomas Collins, who expected to follow some weeks later. The passengers numbered in all about two hundred and twenty men. This number included three engineer corps under Charles M. Bird, as chief engineer, and Charles W. Buchholz, John Runk, and Amos Stiles, his principal assistants.

All these men had seen service in either the army or navy during the Civil War and had held positions of great responsibility on railroad engineering work in various parts of the United States. Many of the men holding positions of secondary importance, such as C. S. d'Invilliers, Joseph Byers, R. H. Bruce, W. C. Wetherill, C. A. Preston, John B. Dougherty, and others, had previously earned enviable reputations on the leading railroads of the United States and were in every way qualified for any promotion the exigencies of the service might require.

Dr. E. P. Townsend was in charge of the medical department. There was a large force of storekeepers, timekeepers, and clerks. Faithful Irish foremen, who had been with P. and T. Collins for years, were still following the fortunes of that firm on board the *Mercedita*. Carpenters, mechanics, and a large party of axmen from the lumber regions of Pennsylvania made up the remainder of the party.

Besides personal baggage, the cargo consisted of five hundred tons of railroad iron; two hundred tons of instruments, tools, general merchandise, and provisions; and three hundred and fifty tons of coal.

An equal quantity of coal had previously been sent by a sailing vessel to Pará for use on the return trip.

The general interest in and the incidents attending the departure of this the pioneer vessel of the expedition were thus described in the Philadelphia *Times* of January 3, 1878:

> The steamship *Mercedita*, Captain Jackaway, at 1 o'clock yesterday cast off her lines from Willow Street wharf and started on her long voyage to the heart of the South American Continent. Since the days of the California gold fever, when the departure of every vessel laden with passengers seeking their fortunes in the new El Dorado was the occasion of an excitement that stirred to the center the sleepy Philadelphia of that time, there have been few such scenes witnessed upon the Delaware front as was shown at Willow Street wharf yesterday. It was not the gathering of the friends and relatives of the two hundred and twenty-seven souls on board, attracting a gaping crowd of idle sightseers, but the profound interest in the departure of the pioneer vessel of a great Philadelphia enterprise, that filled the wharves and shipping for a block with densely packed humanity, anxiously awaiting the moment when the steamship, with her precious freight, should start upon

her journey. Although the head of the Willow Street wharf was roped off and guarded by policemen, yet the crowd at the end was so close that it was with the greatest difficulty that the side of the steamer could be reached. At the gangway stood two of the company's officers, ticking off the names of the laborers as they came on board. The decks were in great confusion, the vessel having been loaded with so much dispatch that it had been found impossible to get her ship-shape before sailing, and up to the last moment a whip, rigged over the forehatch, was kept busy in hoisting in baggage and stores.

As the head of the steamship swung off in the stream the people on shore began to hurrah, the steam-whistles of a dozen or more tugs on the river and a locomotive on the wharf to blow, and until she had steamed out of hearing the cheers from thousands of throats continued.

The police boat *William S. Stokley*, on board of which a number of ladies and gentlemen had embarked for the purpose of seeing the good ship off, kept company until the Old Navy-Yard was passed, when the *Stokley* steamed rapidly ahead. All along the river front the *Mercedita* was saluted by the cheers of the people assembled upon the wharves, by the blowing of whistles and the ringing of steamboat bells. The passengers appeared to be equally enthusiastic, and hallooed themselves hoarse in answer. Off Chester the *Stokley*, which had run a mile ahead, slowed up until the *Mercedita* overtook her, when the *Stokley* ran alongside and Messrs. P. & T. Collins and several other gentlemen (including Colonel Church), who had come thus far on board the steamship, came on board the police boat and a number of the passengers on the latter were transferred to the *Mercedita*. The vessels then separated; there was cheering and waving of handkerchiefs and hearty good-bys, the *Stokley* turned her head towards the city, and the *Mercedita* sped away towards the sea.

Farewells over, attention turned to bringing order out of the general confusion that had prevailed all over the vessel. Many stores required for immediate use had been placed under tons of heavy material in the hold, where they could be reached—with difficulty. The roll was called twice during the afternoon; staterooms were assigned to the few who were luckily entitled to them, about forty in all. Owing to a heavy southeast gale we came to anchor about forty-five miles from the capes of the Delaware. As soon as the engines stopped, the vessel began to give evidence of a capacity for rolling, which was abundantly proved on many subsequent occasions. Even at this early stage in our progress, not a few of the passengers had occasion to seek the ship's side or more secluded places below. Those who were able to eat found great difficulty in securing a cup of coffee and a few sea biscuits. The ship's cook and galley were entirely unequal to the task of preparing food for the number on board.

Rude temporary bunks had been erected between decks forward for the use of laborers, while those in the cabin not so fortunate as to secure staterooms had to be contented with a mattress, pillow, and blanket on the cabin floor or on the cabin table. Those who lingered on deck late that night found it almost impossible to reach their staterooms without stepping on the sleepers, who, as closely packed as a layer of sardines in a box, occupied every available foot of floor space in the cabin. One or two persons had provided themselves with hammocks and in these swung over the heads of their less-thoughtful companions and escaped in a great measure the disturbing effect produced by the rolling of the ship.

Though Captain Jackaway was nominally in command of the vessel, the real captain, except in matters directly relating to navigation, was Gertie, the stewardess, a young woman of not very prepossessing appearance, who domineered alike over passengers and crew from stem to stern of the boat. Her long association with sailors had destroyed every vestige of those traits we are accustomed to associate with women. At any time of the day or night, the strident tones of her voice might be heard reproving those who were so unfortunate as to transgress the rules and regulations she had laid down. When excited to anger, her vocabulary of Billingsgate and rough Irish wit were more than equal to an encounter with the toughest characters on board. In a moment of anger, one of the engineers gave her a nickname, which, though rhyming perfectly with the name she desired to sail under, gave her mortal offense and led to a doubly restricted diet for the offender during the remainder of the voyage.

When, as was frequently the case, she was called from her slumbers at night, Gertie never deemed any elaborate toilet necessary but emerged from her stateroom costumed in a manner, as FitzGreene Halleck would express it, somewhat

"Eve-like, angel-like, and interesting," and caused many an attack of nightmare by tramping barefooted, without concern or apology, lantern in hand, on and over the prostrate forms of those sleeping on the cabin floor.

With all her failings, however, Gertie was kindhearted and often during the voyage, when the regular meals were unendurable, for a valuable consideration, would secure us surreptitiously little delicacies, such as a piece of pie or cake, a can of peaches, or a glass of ice water, probably abstracted from Captain Jackaway's private stock of provisions.

In our attempt to weigh anchor the next morning (January 3), the windlass broke, and it was nearly noon before the vessel was again steaming down the Delaware. Want of organization was painfully apparent. We were divided into messes, apparently by lot. Men were detailed from each mess to wash the dishes and convey the food from the forecastle, where extra cooking-stoves and cooks had been installed, along the side of the ship and downstairs to the cabin table. The newly impressed cooks proved inefficient. Saltwater soap had either been forgotten or was stored where it was inaccessible. Ordinary soap was useless, and the men had not yet learned the trick of subjecting the greasy dishes to a jet of steam. The improvised waiters, not having acquired their sea legs, later proved unequal to the task of carrying the food to the cabin table, and the liquid or semiliquid articles generally reached the foot of the cabin stairs considerably in advance of those persons who were intended to carry them and not infrequently fell upon the heads of the hungry crowd below. The number of messes, which in succession occupied the cabin tables, was so large that meals were being served at all hours of the day and far into the night.

The food consisted at first of coffee without milk or sugar, sea biscuits, and moderately palatable soup.

Later, we were initiated into the mysteries of *plum duff* and *souse*. The first of these well-known and delectable marine delicacies was made in bags from pieces of muslin taken from flour bags or abstracted from the clothesline.

The food was served from large tin washbasins or dishpans, and the coffee in ordinary wooden buckets. Tin cups and plates, with an inadequate supply of knives, forks, and spoons, constituted the tableware. No attempt was made to seat the messes at a table, but all stood around frantically endeavoring to secure and swallow such articles of food as circumstances, including the, at times, awful rolling of the vessel, permitted.

The fresh-water supply was so limited that we were only allowed to use it for cooking and drinking—and then, only in small quantity. Even this was none of the best and tainted both as to taste and color, due either to the tank in which it was stored being unclean or to the alleged fact that it had been taken from the Delaware below where a city sewer discharged itself in that stream.

At 4:00 p.m. we passed the Delaware breakwater and headed southward for the open sea. Our experience thus far had not been encouraging, but the youngsters, then leaving, for the first time, homes where they had been accustomed to every luxury, no less than their elders, who had for years been accustomed to the hardships of camp and field, were disposed to be cheerful under conditions that all regarded as temporary and the inevitable result of the excitement and confusion attending our hurried departure.

The sky was clear, the sea calm, and there seemed to be every prospect of a pleasant voyage. With high hopes and undiminished enthusiasm, we watched the icebound shores of our native land disappear from view and sought our beds that night confident that in a few days, we would be enjoying the balmy breezes of southern seas and basking in the golden sunlight of everlasting summer. About midnight, we were startled by a sudden change in the weather, the rattling of cordage on deck, the taking in of sails, the continuous tramp of sailors overhead, loud orders issued by the captain, and the responsive, "Aye, aye, sir," of the men. Everything seemed to be tumbling about on deck and in the cabin. Crash after crash followed, as improperly stored tin pans and dishes fell to the floor. The vessel was pitching and rolling in a fearful way, and a howling gale from the south made it nearly impossible to hold our course or make any material progress. Even in a moderate breeze, the *Mercedita* had shown a tendency to roll that made her the subject of many a jibe and jest; "She would roll in a canal," "There's a fish under her stern," and similar comments had been frequently heard. This marked peculiarity was now aggravated until it became truly alarming, and all day January 4, many anxious faces could be seen on board. All but a very few of the passengers were seasick; the whole cabin was consequently in a most disgusting condition. Though the rain fell in torrents, some of those who were well enough, with rubber hats and coats, braved the fury of the storm on deck rather than endure the worse condition of affairs below. Throughout the day, it was necessary for those on deck to keep fast hold of some fixture at all times, not only to prevent being washed overboard but also because the inclination of the deck was often such, both sideways and fore and aft, that a passenger not holding fast to something would run an involuntary footrace down the steep sloping deck, and it then became an open question whether he could catch the low railing of the vessel on the fly or would go right over it into the sea.[9]

No regular meals were served that day. The improvised cooks and waiters were, as a rule, somewhat indisposed and had other and more pressing duties to perform than that of satisfying the "aching void" of their fellow passengers. The few who made an attempt to convey soup and coffee along the ship's side and down to the cabin table attained only partial success. That portion of those liquids they succeeded in getting to the head of the cabin stairway was mainly used to deluge the seasick passengers who occupied the cabin floor. It may well be imagined that the remarks that greeted this performance were not such as to encourage a repetition. If any, at times, were inclined to laugh at the sufferings of our more unfortunate companions, the distribution of life preservers, provisioning of the lifeboats, and a rumor that circulated over the vessel to the effect that three New York insurance companies had declined to take a risk on the ship and cargo, brought us to a realizing sense of the serious condition of affairs that confronted all.

With the perversity of fate, however, the ludicrous will at times obtrude itself in our most serious moments, and that storm will ever be associated in the minds of many with a picture of old John O'Hara, the general foreman for P.

---

[9] The New York newspapers of January 6 mention eight wrecks along the New England coast on the fourth.

and T. Collins, a devout Roman Catholic and a man whose courage was beyond all question, as he reclined in the cabin, surrounded by suffering and groaning humanity, like himself, in the worst stage of *mal de mer*. The expression of his face suggested the unuttered prayer, *"Now God be good to me in this wild pilgrimage. All hope in human aid I cast behind me,"* as he devoted his time about equally to relieving his stomach and conscience and sprinkled himself and those about him from a bottle of holy water he had previously provided for just such emergencies.

The few who were able to eat that day obtained with difficulty a little coffee and a few sea biscuits. No arrangements had yet been made for a regular distribution of drinking water, and many suffered for lack of that essential article. We all knew that the ship was making practically no headway and that the obscuration of sun and stars made the usual observations for latitude and longitude impossible. It was about all we could do to keep away from the dangerous coast of Hatteras, and Captain Jackaway would turn with a look of supreme disgust from anyone who had the temerity to inquire as to our day's run or present location.

There was little sleep for anyone on board that night, and those occupying upper bunks in the staterooms had to take every precaution to prevent being thrown to the floor.

All the following day, January 5, the gale continued, but the wind was steadier and had worked around so that it now struck us from the stern. Food was still miserable and hard to obtain. Even as the storm moderated, the tremendous swell it had raised ran high. Every time we crossed one of these mountainous swells and the bow of the vessel took a plunge into the abyss ahead, we thought of the tons of railroad iron in the hold below and wondered whether she would ever emerge again or whether some heavy articles, becoming loose, would knock a hole in the ship below the waterline. At the mess table, as we stood up to eat, a lurch of the vessel would frequently dash the soup or coffee a man was vainly endeavoring to swallow across the table and over someone on the opposite side. A large part of the food, in one way or another, eventually reached the cabin floor, the condition of which can be better imagined than described.

January 6 was our first Sunday at sea. The storm had entirely abated, the sky was clear, and the temperature so mild that almost all were able to enjoy the day on deck. After the dangers we had passed through, there were few on board not disposed to join heartily in the simple morning service that was conducted upon the open deck, peculiarly impressive as it was, being confined by the formalities of no religion and no creed. Vocal and various kinds of instrumental music constituted a prominent feature in the exercises and even yet the familiar lines, "There's a land that is fairer than day, and by faith we can see it afar," are always associated in the writer's mind with that memorable occasion.

The food and manner of serving it continued to be equally disgusting. About this time, many of the men began making private arrangements with the stewardess and cabin boy to supply their wants. During the day, as nearly all the men had recovered from the effects of the storm and had acquired their sea legs, a feeling of gaiety began to pervade the vessel. Singing and music of every description could be heard from bow to stern of the *Mercedita*. The cornet, accordion, flute, hymns, and ribald songs were to be heard on the decks and in the cabin.

The next day, Monday, January 7, we made favorable progress under steam and sail. The meals had slightly improved and, by comparison with previous days, seemed passably good. Pork, coffee dipped out of wooden buckets, apple butter, tomatoes, and sea biscuits seemed to constitute a very fair menu, but many found difficulty in swallowing these articles out of half-washed dishes, and it was difficult for those who could eat to get a bite of the more desirable articles, so quickly were they consumed. Canteens were distributed, and each man was allowed one quart of drinking water per day. A large part of this was used to wash in, as we could not keep ourselves clean with salt water and ordinary toilet soap.

By noon on January 8, we were eight hundred and forty miles in a straight line from Cape Henlopen. That afternoon, the wind freshened until by night it was quite violent. The vessel pitched so that nothing would stay on the cabin table that was not fastened there. The intense heat during quiet weather had caused the woodwork above the waterline to shrink, and the heavy sea reaching these parts caused a troublesome but not dangerous leakage. The cabin and some of the staterooms were afloat, and water from the decks dropped down onto the beds. It became difficult to stand anywhere without holding fast to some fixed object.

At night, one's heels would be alternately elevated and depressed several feet above or below his other extremities. Sleep was out of the question. The storm kept up with unabated fury until the morning of the ninth, when, for a few hours, it ceased entirely. However, in the afternoon, it resumed blowing as hard as ever. Some of the men in the forward part of the vessel became so alarmed that they wore their life preservers during the day and night. Thus equipped, they desired to promenade the decks but were not permitted to do so for fear of exciting the same feeling of alarm in others. That night, the storm became more violent and the sea more turbulent. The writer's diary says, "It is nearly impossible to swallow any liquid substance while crowded together at meals. We pitch against each other and over tables and benches from one end of the cabin to the other."

The heavy adverse wind continued until early on the morning of the tenth and had it not been for the confidence in the vessel our previous experience had given rise to, we should have regarded our situation as dangerous in the extreme. By daylight on the tenth, the wind had entirely subsided, but the heavy swell resulting from it continued, and we all felt that to those enamored of, "A life on the ocean wave, a home on the rolling deep," we could confidently recommend the *Mercedita*.

This day was generally known in the annals of the expedition as "Mush Day." Mush had been prepared in tin pans for breakfast, but a great part of it never reached the table. There was an almost continuous trail of mush from galley to cabin, which rendered it more than ordinarily difficult for the passengers to maintain an erect position while thrown about by the lurching of the vessel like dice in a box.

Captain Stiles, a man of venerable and patriarchal appearance, whom a stranger, at first glance, would have taken for an itinerant Methodist minister rather than a civil engineer, by first depositing his ration of mush on the floor and then falling into it, had apparently performed the remarkable gymnastic feat of placing the entire contents of his pan upon his own back. Then, while the air was sulfurous with the victim's maledictions, Mr. Buchholz appeared upon the scene and, in tones of mild remonstrance, told the gallant captain that he *ought to be more careful!*

By this time, it had become too warm for comfort in the cabin, and on the eleventh, the change in temperature was so decided that summer clothes and straw hats made their appearance. The writer's diary continues to record the fact that "the meals were simply villainous, and complaints without end were heard." The not-very-appetizing rumor circulated through the vessel that the wooden buckets in which coffee was served were not used exclusively for that purpose.

On the twelfth, the weather was so warm and sultry that, for the first time, an awning was raised on deck. Many men were becoming weak and debilitated on account of the diet.

On the thirteenth, a Sunday, the unsystematic individual bribing of the stewardess and cabin boy was improved upon by the formation of a private mess, which throughout the remainder of the voyage became notable, or notorious, as "The Big Twelve." It was composed largely of kindred spirits among the engineers, who had known each other in Philadelphia, but the essential requisite to admission was good fellowship rather than rank or previous condition of servitude. The writer has no desire to injure the well-earned reputations some of this gang have since acquired, so he omits to mention names. Their principal asset, however, was a corner on the services of Gertie, the stewardess, which this organization attempted to monopolize.

About this time, it became generally known that we would have to stop at Barbados for coal, as the supply on board was nearly exhausted. The wind and weather for the next four days continued to be favorable. Games, music, dancing, and storytelling helped to pass away the daylight and evening hours.

At 3:00 p.m. on the sixteenth, the lookout at the masthead created great excitement by announcing land in sight, and before dark, at reduced speed, we were coasting slowly along the shore of Barbados looking for a pilot to come out and take the vessel in charge.

VI. Barbados

There with a light and easy motion,
The fan-coral sweeps through the clear, deep sea;
And the yellow and scarlet tufts of the ocean
Are bending like corn on the upland lea:
And life in rare and beautiful forms,
Is sporting amid those bowers of stone,
And is safe, when the wrathful spirit of storms
Has made the top of the wave his own.

—J. G. Percival

The sun was well up the next morning, January 17, when we were awakened by a most terrific din and clatter of unfamiliar voices, as it seemed right under our heads, the clash of oars and the noise of small boats colliding with each other and with the sides of the vessel. Through the open deadlights, we were not long in discovering that the steamer was slowly moving into the harbor of Bridgetown, pursued by an innumerable fleet of small boats, manned by darkies, yelling, cursing, and fighting worse than a set of old-time New York hack drivers on land and fully impressed with the idea that in the *Mercedita*, with its two hundred and twenty passengers, they had struck a "big bonanza."

From the deck, after coming to anchor, the view was positively enchanting, and it was difficult to realize that two weeks before, we had been shivering in heavy overcoats. From the vessel, we could see the tall coconut palms and many other, to us, unfamiliar specimens of tropical vegetation. At the side of the vessel, we were besieged by natives offering every kind of fruit and propositions to take us ashore. Darkies amused us by diving to an incredible depth after small coins we tossed into the clear water. After two weeks of semistarvation, it can easily be imagined that the quantity of tropical fruit consumed was astonishing. It had been doubtful whether we would be permitted to go ashore, but a few of the passengers settled that question for themselves by swinging off the vessel by ropes into the small boats waiting below. Seeing the uselessness of any attempt to restrict us, the officers of the vessel later lowered the landing stage, and all went ashore as fast as the small boats could carry them. So effectually had we acquired our sea legs that it was laughable to see the men on their way from the wharf rolling like sailors from side to side.

The first thing that impressed a stranger in the Bridgetown of that time was the fact that a large majority of the population consisted of darkies. The streets were nearly all curved or made up of frequently broken lines. They were very narrow, and only the more important had sidewalks, varying from eighteen inches to two feet in width. The walls of the houses were nearly always built close to the curb lines so that their eaves overhung the sidewalks. One of the striking peculiarities of the island is that it is a coral formation and not of volcanic origin. This substance furnished a natural pavement for the streets as smooth, hard, and durable as asphalt and a satisfactory building-material for dwellings of the better class. Most of the dwellings of the darkies were made of wood. We saw no brick on the island. The temperature was delightful and varied very little the whole year round. The thermometer in summer, we were told, seldom rose above 80° Fahrenheit in the shade. The darkies could frequently be seen riding in carts drawn by little donkeys about the size of a two-month-old calf. One of these little animals would often draw as many as four persons through the streets at a brisk trot.

The island is about twenty miles long and fourteen wide, with somewhat the shape of a ham. It was said to have a population of 165,000, of which 13,000 were whites. English was supposed to be the language spoken, but isolation had produced among the darkies a dialect that it was often difficult for an Englishman or American to understand. The darkies made a very good appearance in their straw hats and clean white suits but were worse than Jews at driving a bargain. They would agree to take passengers ashore for a shilling each but demand double that amount when they reached the wharf. Unless we became as frantic as maniacs and threatened to club or shoot them, they would follow us all over town in hopes of obtaining a sixpence for some service that was merely a legal fiction.

Two of our engineers inquired about the best hotel and were recommended to the Albion House. After they had vainly endeavored to discover it themselves and having been informed that it was five miles out of town, they engaged a cab to take them there. After about a two-hour ride through the country, which, by the way, was so attractive that the time passed quickly, the driver finally brought them up in front of the hotel, which they subsequently found to be only two minutes' walk from the starting point. Though not a pretentious place and not the best hotel in Bridgetown, the Albion was to us at the time like a miniature Waldorf-Astoria. To be able to sit down at a clean table to an excellent meal consisting of mutton chops, coffee, sliced cucumbers, bananas, and oranges, followed by English ale and choice Havana cigars and to be able to repose at night in a clean and immovable bed seemed the *ne plus ultra* of human happiness. Much of our time was employed in exploring and sightseeing.

We were greatly interested in the sugar plantations, which covered the island and in the great number and variety of fruit and shade trees, many of which were entirely unknown to us. Indeed, it has often since occurred to the writer that he encountered a greater variety of tropical productions in Barbados than during the whole trip up the Amazon and Madeira. Tall coconut palms with their tempting fruit, furnishing a delightful draft of clear and cool water in the warmest weather; bread-fruit; honey locusts; mangoes; limes; lemons; oranges; pineapples; and other such products, together with the luxuriant foliage and delightful sea breezes, made the transformation scene that had occurred within the brief period of two weeks seem like the masterwork of some great magician.

At the Ice-House, a mixed affair where hotel, restaurant, and wholesale and retail fruit and provision store were all combined under one roof, we found a large party from the *Mercedita* and many of our engineers, all apparently engrossed with the one idea of atoning as far as possible for sins of omission in the past and fortifying themselves to resist the effects of similar evils in future. Not a few of our party were what another has aptly characterized as "men with the bark on," relics of a generation now nearly extinct, who had passed through the trying times of the Civil War and gained their engineering experience on the Pacific railroads of the West at a time when the typical engineer of the day was a combination of railroad man, Indian fighter, and cowboy. It will not, therefore, be surprising that the truthful historian feels compelled to record the fact that some few of the men he encountered at the Ice-House were in a decidedly hilarious mental condition and though "half-seas over" were still calling upon the landlord to "fill the flowing bowl."

Not a few of us had received invitations to attend a *dignity ball* that night, an institution peculiar to Barbados. Unfortunately, the writer was unable to attend, but so far as he could judge from the description of others, the name was a *lucus a non lucendo* in derivation.

At a place known as "Hastings," about three miles from town, where a large modern hotel has since been erected, many of us indulged in a luxurious sea-bath, between the main shore of the island and the coral reef that encircles it, at an expense, for towels and soap, of eight pence each.

On the morning of the eighteenth, the guests at the Albion were awakened by the landlord placing a small table at the side of each bed, upon each of which were a cup of coffee, a piece of toast, a plate of fruit, and some fine cigars. Seeing that the writer was awake, he remarked, "Breakfast at nine, sir," and disappeared. Our time that day was about equally divided between driving over the island, eating, and bathing. Some of the men, as a result of the previous evening's festivities, were still suffering from slight mental aberration, but only three had gone so far as to require someone to take care of them. Two, who started in a small boat to the *Mercedita*, were in such a condition that when one fell overboard, the other had not strength to pull him back into the boat, and the oarsman was afraid to interfere lest, in a contest with the two, the boat would be overturned. He therefore continued on his way, while his only remaining passenger, holding fast to his companion's collar, towed him astern. The fellow in the water was comparatively sober by the time he arrived at the vessel, but the other had no sooner reached the ship's cabin than he began to celebrate his achievement by firing his forty-five Colt through an open deadlight in such an indiscriminate way as nearly to frighten the wits out of the darkies in the numerous boats alongside. Fortunately, this amusement was interrupted before any serious damage resulted.

At the Ice-House, guests enjoyed what to them was a rare treat—green-turtle soup and a plate of flying fish. There were stationed on the island quite a large number of English troops, many of whom had served in the Ashantee Campaign, and in the evening, some of us drove out to the garrison to hear an open-air concert by a military band of forty pieces. Previous drives had proved so enjoyable that they were difficult to improve upon, but the light of the full moon, the luxuriant foliage, the brilliant uniforms, the balmy breezes, and martial airs of England all combined to produce an effect that can be felt but not described.

We had been notified to be on board at an early hour that

evening, but a party of us felt perfectly safe in lingering on shore until nearly 10:00 p.m., because one, more thoughtful than the others, finding Captain Jackaway in a hopeless and helpless state of intoxication, had put him under lock and key at a convenient resort near the wharf. After hiring a large six-oared boat and depositing Captain Jackaway safely on board, we were soon on our way to the steamer. The darkie oarsmen had good voices and seemed to be provided with a repertory of patriotic songs calculated to draw generous compensation from anyone, whatever his nationality, who might chance to fall a victim to their wiles.

On the way, they gave us "Marching through Georgia," and as we neared the *Mercedita*, another song, apparently in part of their own composition, with the chorus:

> *Then hurrah, hurrah for the bonny blue flag,*
> *The flag that set us free.*

At eleven that night, we weighed anchor and with many a sigh bade farewell to Barbados. Our stay could hardly have been more agreeable. The writer has often wondered since whether the place was really as attractive as it seemed or whether the impression made upon his mind was due, in a great measure, to the contrast with our previous privations. Certain it is that the effect produced upon us could be compared with nothing else than the vision that burst upon the enraptured gaze of the wondering Peri through the opening gates of Paradise, and Bridgetown, with its diving darkies, donkey carts, green-turtle soup, flying fish, fruits, and flowers will long be remembered by us all.

VII. Voyage of the *Mercedita* Continued

Silent we sailed the phosphorescent seas,
Our ship a craft with shadowy masts and spars,
While gloomed above, in glorious galaxies,
The phosphorescent stars.
—Clinton Scollard

The trip from Barbados to Pará furnished few incidents worth noting. The voyage would have been delightful had not the meals continued to be an unceasing cause of irritation and complaint. The weather was so warm and pleasant that nearly all the passengers carried their mattresses on deck and slept there at night. Schools of flying fish by day and the phosphorescent waters by night were the only novelties to attract our attention.

There were a few cases of sickness on board after leaving Barbados, but none of a serious character. Frequent showers became a noticeable feature of the region through which we were passing. These would come up with remarkable suddenness, and in a very few minutes, a perfectly clear sky would be entirely obscured by dark clouds. A heavy downpour of rain would follow, and almost before we could obtain shelter, it would be clear again, without the minutest particle of cloud in sight to indicate the possibility of recurrence. Yet, sometimes, this would be repeated at frequent intervals all day and kept us constantly changing from deck to cabin and vice versa, as circumstances might require.

The men were expected to take care of and be responsible for mattresses used on the cabin floor or carried to the deck at night. When anyone failed to find his own, he generally squared accounts by appropriating the first he could lay hands on. Fred Lorenz, a member of the engineer corps and a general favorite, had landed in the United States from Germany only a short time before joining the expedition and was, therefore, about equally well versed in the English language and American morality. When he failed to find his own mattress, a nightly occurrence, he regularly reported the fact and drew another from the cabin boy. The result was that near the end of the ocean voyage, he was charged with having *consumed* twenty-five mattresses, which, to his benevolent protestations as to the impossibility of any such occurrence, he was further comforted with the explanation that the mattresses had probably been left on deck and washed overboard.

Our apparent day's run would vary from 120 to 155 miles by the log, but this was against a strong current and greatly exceeded our actual speed.

The temperature in the cabin at noon was generally close to 82° Fahrenheit.

On the twenty-fifth, great consternation prevailed among the cabin passengers in consequence of the discovery that some of the laborers in the forward part of the ship were infested with that species of vermin an individual member of which the poet Burns had immortalized in an epigram intimating that its essentially disagreeable character was in no way modified by association with royalty, not even "Though it crawl on the curls of a queen."

The result was a rigid quarantine, which prevented all communication between the cabin and other passengers.

By noon on the twenty-sixth, we were in north latitude 3° 24', and that night, for the last time on the voyage, we saw Polaris. The pointers seemed almost to speak, as, from a few degrees' greater elevation, they directed our attention to this friend, who had presided over our destinies from childhood, now preparing to yield our future guidance to the Southern Cross and sink beneath the waves.

Early on the twenty-seventh, we began to notice the decreased transparency and change in color of the water. The muddy-green color became darker as we proceeded, and, though far out of sight of land, this fact proved to us that we were already sailing through the waters of that vast but comparatively unknown, empire—"where, amid a pathless world of wood, gathering a thousand rivers on his way, Huge Orellana rolls his affluent flood."

At 4:00 p.m. on the twenty-eighth, we sighted land. At 7:20, one of the mates, who, under orders, was steering a compass course without a lookout on deck, nearly ran down the lightship at the mouth of the Pará River. We were only saved by Captain Symmes, one of the passengers, seizing the wheel and throwing the vessel off her course. As it was, we had a narrow escape and passed so close to the lightship that we could easily have jumped on board of her. At 9:30, we dropped anchor in the rapid tidal current of the Pará

River. In less than a minute, six hundred feet of chain cable rattled overboard never to be seen again. At 10:30, passengers discovered, before the ship's officers, that we were going, head-on, straight for the Braganza shoals, the most dangerous on the coast. The breakers were in plain sight, and their deafening roar could be easily heard as the vessel came about. Soundings showed only nine feet of water under her keel. We then put to sea and beat all night about the lightship.

Fortunately, only a gentle breeze was blowing. At midnight, a fire broke out in the forecastle, caused by the overheated chimney of a stove, and was, with no little difficulty, extinguished. Later, the lantern at the masthead broke or exploded and, with its blazing contents, fell crashing to the deck.

At 9:00 a.m., on the morning of the twenty-ninth, we secured a pilot and proceeded on our way up the Pará River. The run of about 115 miles that day convinced us that the *Mercedita* would be almost unendurable as a habitation during the remainder of the journey. We could not sleep on the unprotected deck at night in the middle of the rainy season, and the heat of the cabin between decks, now that we had ceased to feel the bracing sea air, was absolutely beyond endurance.

Many of the boiler tubes were said to be burned out.

The journals of the engine were reported to be loose and badly in need of repairs. The experience of the past night had led us to contemplate a boiler explosion as among the extremely probable events of the day. It was, therefore, with a feeling of intense relief that at 7:30 p.m., we dropped our only remaining anchor off the city of Pará.

VIII. Pará

*Between the palms and the sea,
Where the world-end steamers wait.*
—Kipling

It is not the writer's intention to emulate the example of a famous newspaper correspondent, who is said to have mailed, on the morning of his first arrival in Paris, the best description of that city ever written, one that he had prepared *en route* from guidebooks and accounts of previous travelers.

To those who desire a minute description of places and scenes along the Amazon and Madeira, there is a wealth of information accessible, gathered by our many illustrious predecessors and successors.

It is to be understood that our one great aim was to reach San Antonio and begin work within the time specified by contract. All other considerations were of secondary importance. The hasty and somewhat superficial glance we were able to obtain of objects in passing may, nevertheless, convey a more correct impression to the mind of the reader than a more elaborate picture confused by a too-great mass of petty detail.

On the morning of January 30, as we came on deck and obtained our first view of the city and Harbor of Pará, it must be confessed that the general impression was one of disappointment. Previous travelers had led us to expect much in this, the great commercial metropolis of the Amazon. The large number of vessels that floated nearby, bearing the flags of nearly all the maritime nations of Europe, it is true, indicated a considerable degree of commercial activity, but the general appearance of the city differed little from that of any old Spanish town and even a more pretentious place would have suffered greatly in appearance by reason of the flatness of the picture and the dense mass of green tropical foliage surrounding it in every direction except where broken by the turbid waters of the rapidly flowing river.

From the vessel, we could see the houses, many of which were dingy and brown from age. They were generally only two stories in height. Some of them had fronts made up of little porcelain squares of various colors. The roofs of the buildings were quite conspicuous from the red brick semicylindrical tiles of which they were composed. Such roofs, though by no means ornamental, were well adapted to the climate, entirely excluding sun and rain while permitting a free circulation of the air. Covering the roofs of the houses and strutting about the streets and yards were to be seen great numbers of hideous buzzards, those omnipresent scavengers of the city and its only protection against pestilence. Apparently aware of the protection afforded them by law, they were far less easily frightened than chickens in a farmyard.

Going ashore for a closer inspection, we were impressed with the strength of the current, due to a tide that here attains a height of about fifteen feet. The ship's cutter, with six hardy sailors at the oars, was swept half a mile out of its course before we reached the wharf. Our party divided themselves between the two best hotels, the Hotel Central and the Hotel do Commercio, and from these centers separated into smaller groups for purposes of exploration.

Before our arrival, the emperor, recollecting the hospitality extended to him while in the United States, had issued an order commanding all officers and employees of the government to render us every courtesy and assistance in their power. Few of the passengers discovered this fact, owing to want of familiarity with Portuguese, but those of us who did made good use of it. A few were so fortunate as to fall under the guidance of a gentleman, whose visiting card is still in the writer's keeping and reads:

O Commendador
M. A. Pimenta Bueno.

Mr. Bueno had, we understood, held high official position under the government and at that time was president of the Amazon Navigation Company. He was a man of highly polished manners and fine education and was able to converse fluently in any one of several languages. We were entertained most hospitably at his suburban home and were greatly assisted by his advice in our search for places of interest. The writer particularly recollects him, however, in connection with a present of the choicest Bolivian tobacco, the receipt of which caused a large stock of the North American article to be consigned to the river the next day.

Among many other places, we visited the Fort of Pará, where the government maintained a school for training soldiers in mechanics and music. We were there introduced to the commandant, a professor of geometry, and the director of the arsenal. The professor spoke English quite well but occasionally made slight mistakes.

While inspecting the fort, he took us into the hospital and with a sweep of his hand indicating about forty invalids lying on cots in a very emaciated condition, said, *"These all have yellow fever."* For a moment, we were dumb with surprise and astonishment. The same thought seemed to run through all our heads simultaneously—that it would be almost criminal for us to return to the *Mercedita*. The professor suspected from our looks that he had made a mistake and quickly explained that he had intended to say *malarial* fever.

A little later, some of us visited the office of the leading newspaper, the *Diario do Gram-Pará*. The editor, Sr. José Palinoda Silva, was exceedingly courteous, and at his urgent solicitation, one of the party wrote a short account of our voyage, which, after translation, probably appeared in the next issue of the paper.

Professor Orton, who visited Pará in 1868, gave the population as thirty-five thousand, and, as it had doubled in the previous twenty years, it is fair to presume that at the date of our visit it fell little, if any, short of forty-five thousand. This population was composed of individuals claiming nearly every nationality on the face of the earth, and, as in Brazil there is little of the prejudice existing here against amalgamation of the white and colored races, we found many specimens having features of a composite character that, under skins of every hue and color indicated a mixed extraction.

On the wharves, Negroes performed the more laborious duties, but many of the boatmen seemed to be representatives of Portugal. Negro and Indian women could be frequently seen on the streets, loosely clad in a single white or colored muslin garment cut very *décolleté* and carrying trays of fruit upon their heads. Here and there, nude children capered about us and well-fed priests, under the protective shade of black silk umbrellas, slowly marched along, very conspicuous, however, on account of their black gowns, shovel-shaped hats, and low shoes with immense metal buckles.

The language of the place was as variegated as the population. French, Spanish, and Portuguese would serve a stranger about equally well in intercourse with the natives.

Fred Lorenz, who had been struggling during the voyage to obtain a working knowledge of English, on being confronted with so many other languages, remarked, in a sad and discouraged tone of voice, "I ken not larn so meny dem languages." At a store, the writer attempted to obtain an article of such peculiar character that a very intimate knowledge of at least one of the previously mentioned languages was required to describe it. After a vain attempt to make himself understood in French and Portuguese, the storekeeper finally asked in Spanish, "What language *do* you speak?" On receiving the information that English was about the only language his customer could speak with the requisite degree of fluency, he replied in English, "Well, go ahead; *I was born in New York.*"

The market, with its wonderful display of tropical fruits,

pottery, and curiosities, was a place of never-failing interest to the stranger. There, for the first time, we tasted the famous bananas and pineapples of Pará. The bananas, about three inches in length and encased in a skin hardly thicker than the kid in a lady's glove, had a flavor unequalled by any other in any part of the globe. The pineapples, when cut, were of a golden-yellow color, and sugar could not in the slightest degree increase their sweetness.

As the natives did not use milk or cream in coffee, there was but a small quantity of the lacteal fluid consumed. That required for culinary use was generally sold in the form of condensed milk, but to feed young children and for some other purposes, a small quantity of the natural product was peddled about the streets by driving the cow from door to door, with small tin pails hanging from her horns, and milking her as the demand might require. The cow was generally followed by a calf, whose age was supposed to indicate the freshness of its mother and which fed upon the surplus not required by customers. There were those, however, who did not hesitate to insinuate that, just as beggars at home borrowed babies to excite sympathy, so the milkmen of Pará did not hesitate to borrow calves as a guarantee of the excellence of their milk.

With singular inappropriateness, the name Nazareth, which we are accustomed to associate with poor and humble surroundings, is in Pará applied to a section of the city inhabited exclusively by the wealthy aristocracy of the place. The houses in this part of the city were large and airy, generally only one-story high but surrounded with wide verandas and spacious grounds filled with tropical fruits, shrubbery, and flowers. The avenues in this section were wide and well shaded by trees but generally unpaved and destitute of sidewalks. Even in the business section, where there were sidewalks, they were extremely narrow, irregular, and poorly laid. Pará had absolutely no sewerage and depended entirely on the buzzards to carry off and consume decaying animal and vegetable refuse.

The Estrada de Palmas was the avenue upon which the inhabitants chiefly and justly prided themselves. It derived its name from the palms that bordered it for half a mile, whose high-spreading branches almost interlaced one another overhead.

Mr. R. H. Hepburn, who saw much of Pará, has this to say in regard to its laundries:

> Their laundry system was one of the most peculiar the mind of genius could have evolved. Not satisfied with pounding the linen between stones in lieu of a washboard, they generally took two or three weeks to return the apparel instructed to their care, in the meantime hiring it out on Sundays and holidays to any needy dago who could pay for it. I always regarded with suspicion every person of dubious countenance ostentatiously displaying white linen, whom I met; for I never knew whether he might not be disporting himself in my shirts.

The numerous churches of Pará were not different from those generally found in Spanish-American countries, almost invariably fronting on the public plazas with gable fronts supported on each side by square towers. The cathedral, the

Episcopal palace, the president's house, and a theater, recently completed, were the most prominent buildings we saw.

The theater received an annual subsidy from the government and seemed to be a very popular institution. At the time we were there, an actress was advertised to appear in a costume made up of feathers dyed in brilliant colors, somewhat similar to the famous Brazilian feather-flowers. It was a subject for regret that we were compelled to leave before the day fixed for this performance.

The custom, which secludes, in a great measure, from public gaze the aristocratic ladies of Pará and prevents their appearance unveiled and unattended on the street and in public places, prevented us from passing judgment on their charms, but some of our men who claimed to have caught shy glances from beneath the lace mantillas of the Brazilian *señoritas* on their way to early Mass or had seen fans and handkerchiefs surreptitiously waved from behind barred and shuttered windows of upper balconies were satisfied that the younger generation was not in entire sympathy with the restraints imposed by Spanish custom.

One institution with whose shortcomings we early became familiar was the post office. Absolutely no order, method, or system prevailed in the handling or delivery of mail. All that class of mail which at home passes through the general delivery window was in Pará deposited on a wide windowsill, where anyone who desired to do so could help himself. After remaining there for a time, the mail not called for was removed to a large table just inside the official railing, where anyone making application might examine it for himself and appropriate such letters or packages as might suit his fancy.

The mail was arranged in piles according to the foreign port from which it came. The only question put to those in search of letters was, "Where do you expect mail from?" and, on giving a satisfactory reply, the applicant was politely directed to the New York pile or the Liverpool pile or whichever one might be desired. The officials seemed to feel that, this done, they had performed all their proper functions. No inquiry as to name or residence was ever made. Not infrequently, valuable letters were carried away by persons to whom they were not addressed but who took them for friends in fear that they might otherwise be lost.

Before the establishment of the Roach Line of steamers, there was no regular mail between New York and Pará. Often afterward, letters mailed in the United States reached us at San Antonio bearing the postmarks of Liverpool, Lisbon, Rio de Janeiro, Pará, and Manáos, and many never reached us at all.

Pará, in our time, had but one street railway. The cars of this line were drawn by mules driven tandem, and the conductor signaled to the driver with a tin trumpet.

While in the city, most of us supplied ourselves with hammocks, which in Brazil are a necessity and not a luxury as with us. They were made for practical use and comfort and could be had in great variety and of most excellent quality.

At the Hotel Central, the servants were nearly all Indians, and Mr. Vierra, a native of Rio de Janeiro, who had joined the expedition in Philadelphia, gave the writer a practical demonstration of the honesty of these people by purposely leaving his pocketbook, containing one hundred and fifty dollars, all day on his room table without losing a cent. In all the time we were in Brazil, the writer does not recall a single case of theft by these domesticated Indians, and months later, while camped on the banks of the Jaci Paraná, when provisions were practically exhausted and he was reserving the very few remaining loaves of bread for the engineers, the *capitão* of the Indian boatmen, who might easily have stolen anything we had at night, selected an hour in broad daylight, when the writer was seated at the camp table, to leave the little Indian camp, ostentatiously abstract from ours a single loaf of bread for his suffering men, and carry it away in plain sight of all.

We were hardly competent, after life on the *Mercedita*, to form a correct opinion as to the merits of the hotel fare. Almost any ordinary meal would have seemed to us at that time to be on a par with the best efforts of noted American hostelries. Many of us, nevertheless, retain very lively recollections of the salads, excellent coffee, and other beverages prepared for us by the obliging landlords, and of instructions received on the proper way in which to cut and eat the delicious little oranges with which the table was supplied.

The currency of the country was based upon the *rey*, which has no more existence in the United States than the *mill* has and was worth only half as much. The smallest banknote was the *milrey*.

As the men received their hotel bills, not a few were astounded at the figures and were greatly relieved when it was explained to them that the totals had to be divided by two thousand to obtain an equivalent in American dollars.

IX. Pará to San Antonio

Where the undethronéd forest's royal tent
Broods with its hush o'er half a continent.
—James Russell Lowell

The *Mercedita* made her first attempt to leave Pará on the evening of January 31 and kept on attempting for several days thereafter without getting out of sight of the city. We became convinced that there was some truth in the well-known proverb that "He who comes to Pará stays there." Three or four times, the *Mercedita* ran aground, and on each occasion, we had to wait for at least one high tide to get her off. This did not agree very well with Professor Orton's statement as to the Harbor of Pará that "ships of any size will float within one hundred and fifty yards of the shore," and a further statement in regard to the Amazon that "for two thousand miles from its mouth, there are not less than seven fathoms of water." The *Mercedita* drew eighteen feet. The writer mentions these mistakes, not for the purpose of discrediting Professor Orton, but for the reason that P. and T. Collins were severely criticized for not availing themselves of information collected by previous explorers. If such errors could be made in regard to one of the principal seaports of Brazil, frequented by the ships of all maritime nations, we naturally argued, what reliance could be placed upon published information regarding the comparatively unknown wilderness beyond!

These experiences were so absolutely contrary to our expectations that not a few persons on board were of the opinion that the pilots furnished us by the Amazon Navigation Company were deliberately running us aground either to compel us to use that line for transportation or to so delay us that we could not reach San Antonio in time to begin work at the date fixed by contract. That this suspicion was unjust was subsequently proved by the fact that the Roach Line steamers, drawing twenty-one feet, could not get up the river nearer than two miles to Pará.

During these four days, it began to appear doubtful whether our engines were sufficiently powerful to stem the swift currents we would have to encounter in some parts of the Amazon and Madeira, then in flood. The time passed by the vessel on sandbars was partially employed by the passengers in visiting rubber grounds nearby and shooting birds of brilliant plumage they found there.

On one of these tramps, we encountered and killed a specimen of the deadly coral snake, of the species Orton describes as having then recently been discovered by Cope and named by him the *Elaps imperator*. While these delays were occurring, another took place that caused us no little anxiety.

Chief Engineer Bird, doubtless impatient at the delay, like the rest of us, and thinking to utilize in some measure the time that would otherwise be entirely wasted, decided to test the capacity of a small steam launch, we had brought for work above San Antonio by running up the river ahead of the vessel until overtaken. Taking with him Captain Symmes, who had had long experience at sea; Frank Snyder, who had taken an engineering course at the Naval Academy; a man by the name of Fisher; and all the coal, provisions, guns, pistols, and ammunition that he deemed necessary, Mr. Bird bade us good-bye and started on his way.

On Monday, February 4, Colonel Jameson, at the solicitation of the engineers, and knowing that, under the most favorable circumstances, we had not much time to spare in reaching San Antonio, decided to charter one of the Amazon Navigation Company's steamers to convey the engineers and a party of axmen with supplies to San Antonio. To effect this much-desired result, the engineers had assigned one hundred dollars each out of their salaries, on condition, however, that if the *Mercedita* eventually failed to reach San Antonio, the money was to be refunded.

Accordingly, eleven of us rowed to Pará in the ship's cutter and quickly concluded a bargain by which we secured the steamer *Arary*, having staterooms sufficient for the entire party and being properly equipped to convey us with comfort to our destination.

Returning on the *Arary* to the point where we had left the *Mercedita* fast on a sandbar, we had her alongside about dark, and then another difficulty confronted us. The laborers who were to remain on the *Mercedita* refused to aid in the transfer of baggage and supplies unless they were permitted to accompany us. There was no other way, therefore, to accomplish the transfer than to organize ourselves into a body of amateur stevedores. Until long after midnight, naked to the waist, we worked as men in that capacity were never known to work before. We rigged our own hoisting tackle and in the stifling atmosphere of the *Mercedita*'s hold, as well as on the decks above, kept boxes, barrels, and baggage moving at such a rapid rate that at 1:00 a.m., with skinned knuckles, exhausted, bruised, and battered, we felt justified in taking a well-earned rest. At 5:00 a.m. on the morning of the fifth, we were at work again and kept it up until noon, when word was passed that the *Mercedita* was afloat and sufficient cargo transferred. About 1:00 p.m., the *Arary* started with the *Mercedita* in tow. We all had supper together that night on the *Arary* and then parted with the *Mercedita* from our comparatively luxurious quarters accepting with equanimity the jibes and jeers of our jealous and less-fortunate companions.

Before leaving the *Mercedita*, Colonel Jameson, who remained on that vessel, ordered Mr. Buchholz to take charge of the party on the *Arary* and, in the absence of Mr. Bird, to act as chief engineer.

During the night, we traversed that little inland sea known as the Bahia de Marajo. About 1:00 a.m. a cry that sounded very much like the English word *fire* was started by the deckhands and reechoed from one end of the steamer to the other. A minute later, we collided with another and much smaller side-wheel steamer coming down the river. Fortunately, the small boat had discovered us in time to swerve from her course, and the bow of the *Arary* did no other damage than to knock a large hole in one of her paddle-boxes. After our experience on the *Mercedita*, such a trifling incident was only sufficient to make some of our men turn over in their hammocks and inquire as to the cause of the disturbance. On receiving an explanation, one of them, not more than half awake, said, "Is that all?" and resumed his slumbers.

As the night was perfectly clear, there was a strong suspicion on the part of some that the collision was not accidental and that the officers of the *Arary* had deliberately intended to sink this rival in the river trade.

Continuing up the Pará River, which gradually narrowed from a width of eight or ten miles, at 1:00 p.m. on the sixth, we stopped for wood at Breves, a little town on the south side of Marajo Island, which exhibited more business activity than any other we had seen above Pará. For hours afterward, we passed through the Estreito do Breves, one of many small waterways connecting the Pará and Amazon Rivers. So narrow was this channel in places that there seemed scarcely room for two steamers to pass, and the branches of the trees on opposite sides almost met overhead. Troops of monkeys were frequently seen, and the water's edge was lined with thousands of a species of worthless waterfowl that, at a little distance, greatly resembled our American partridge. As nearly all the passengers were provided with firearms of some kind, guns, rifles, and pistols kept up an incessant rattle, as we rapidly advanced through this natural canal with its almost solid wall of dense tropical vegetation, towering on both sides to a height of seventy or eighty feet.

At 4:00 p.m., we again stopped for wood and began to realize that these frequent stops were likely to prove an important factor in any correct calculation of the time required to reach our destination.

The *Arary*, going upstream, burned fourteen thousand sticks of wood per day, each stick being one meter in length. In loading the wood, the deckhands and natives on shore formed a line, often of thirty or forty men, extending from the deck to the woodpile on shore. The sticks of wood were then passed, one at a time, along the entire line, while all counted aloud and in unison, "Um, dous, tres ..." Every tenth stick was thrown, as a marker, on a separate pile. It was even alleged that, in case of a mistake in the count, it was the usual practice to pass the wood back and count it over again.

Of course, this method limited the speed of loading to that at which the end man at the woodpile could pick up and hand over the sticks, and as all were very deliberate in their movements, this landing for wood generally involved a delay of several hours, unless the woodpile failed to furnish the requisite supply. To the passengers, these stops were particularly exasperating when they occurred at night, and the monotonous counting of thirty or forty men with the noise of falling sticks of wood made sleep impossible.

The wood stations were usually at the wildest and least important points on the river, the principal requisite being an abundance of large timber and ground on which to store it above the flood level of the river. There was generally little to be seen except a palm-thatched shed to shelter the woodcutters, surrounded by a tropical jungle that, for variety and density, no part of the known world can equal.

On the seventh, at 6:30 a.m., we tied up at another wood station about thirty miles below Gurupa and remained there six hours. At 2:00 p.m. that day, we passed the mouth of the Xingu and entered the Amazon, which at this point, has a width of about ten miles. The apparent expanse of the Amazon is usually disappointing to travelers, for the reason that the river is so filled with islands that often but one of two or more channels is visible.

At daylight on the eighth, we had a distant view of the high, flat-topped mountains of Almeyrim far to the north of us, the first high ground we had seen in Brazil. At 10:00 a.m., we passed Prainha and five miles above stopped a few hundred feet from the north bank at a cattle station. In the warm and moist atmosphere of the Amazon, particularly during the rainy season, it is impossible to keep meat more than a very few hours. It was, therefore, the practice to take on cattle alive and kill them on the lower deck forward at such times as the meat was required.

The process of loading was a novel one to us. The enclosure where the cattle were fenced in had a narrow, funnel-shaped opening terminating in the deep water of the river. Into this narrow passageway, the cattle were driven, one at a time, until compelled to swim. At this stage of the proceedings, two expert Indians in a canoe appeared upon the scene; one paddled, while the other, with a lasso, sat in the stern. The latter quickly threw the lasso over the horns of the swimming animal, and by this means, it was easily guided to the side of the steamer, where, by hoisting tackle attached to the base of the horns, it was quickly lifted onto the steamer's deck with as little apparent effort or injury as would occur in lifting a rabbit by the ears.

At this point, we begin to entertain serious fears as to the fate of Mr. Bird and his companions in the steam launch. It seemed impossible that he could have reached a point farther up the river in the time that had elapsed since his departure. We had maintained a lookout for him on deck night and day. At every stopping place, we had made careful inquiry, but no one had seen anything of the launch. The craft, at that time, was so unusual on the Amazon we thought it could not fail to attract attention in passing. It would have been an easy thing for us to pass the launch without seeing it, either by night or day, in the labyrinth of intricate channels connecting the Pará and Amazon Rivers, of which no existing map gives any idea. Our fear was, however, that in the passage of the great Marajo Bay, often as stormy and dangerous as the open sea, the launch had been swamped and all on board lost.

The only noticeable feature in the vegetation ashore, so far as we could tell from the steamer's deck, was the rapidly decreasing number of large palms, which had been very conspicuous, both as to number and variety, on the river below. Rain was of daily occurrence, and the atmosphere was apparently saturated with moisture. A new pin, carried in the coat all day, at night would not only be completely covered by rust, but so eaten by it that it had the appearance of having been buried in the ground for a hundred years. Even our nickel-plated Smith and Wesson revolvers, half an hour after firing, would be so rusted that the chambers would not revolve until soaked in kerosene oil.

The *Arary* was the best vessel of the Amazon Navigation Company's line and had ample stateroom accommodation for all, but staterooms on these boats were only used as places to store luggage and change clothing.

The upper deck had a solid roof as a protection against sun and rain, and under it, we all slept and had our meals served.

The iron framework of the roof was provided with numerous hooks, on which passengers hung their hammocks. In these, we reclined by day and slept at night.

The hammocks were only removed at mealtime. At seven in the morning, we were served with coffee and crackers. At ten, we had breakfast. Dinner was at one and supper at six. The exact bill of fare at one of the evening repasts was noted at the time as consisting of roast beef, rice, bread, coffee, crackers, pickles, guava jelly, bananas, raisins, figs, and claret. Besides these meals, we were always supplied at three in the afternoon with a large earthenware monkey filled with sweetened water. This practice of drinking sweetened water the writer subsequently found to be customary both in Mexico and the Republic of Colombia.

In the latter country, a peon who could scarcely afford a decent meal would often be found at lunchtime to have a pound lump of brown sugar in his pocket, and in some places, they made a regular practice of putting large quantities of this sugar into the water before permitting the mules to drink. So general was this usage that the particular grade of sugar used was often spoken of as *"dulce para machos"* (candy for mules).

None of the officers or crew on the *Arary* could speak a word of English, and we were early forced to resort to our Portuguese dictionaries for the names of articles served at meals. The laborers, who had provided themselves with no such aids, were a constant source of amusement in their endeavors to make themselves understood at the table. They were striving to get the Portuguese names of articles, and the waiters were equally zealous and generally more successful in acquiring the English nomenclature. The result was a valuable lesson in formation of dialects. The waiters came sufficiently close to *roast beef* to call it "*rois buff.*" The laborers immediately jumped at the conclusion that this was the correct Portuguese name, and thereafter, by common consent, *rois buff* was adopted by both sides, each thoroughly satisfied with their flattering success in mastering the language of the other.

The self-complacency of one native of the Emerald Isle was, however, somewhat disturbed when, after sending for a second plate of *rois buff,* the waiter returned empty-handed, explaining that there was *"nãs mais"* (no more). The words sounded exactly like *no mice* and the enraged Irishman, shaking his fist at the frightened waiter, exclaimed, "The divil take ye! Who asked for mice?"

The waiters at times were none too attentive, and one of the engineers, finding American profanity wasted, made out a complete list of all the profane and opprobrious epithets he would find in the Portuguese dictionary.

The native language is quite well provided with words that are expressive and consign unfortunate humanity to all sorts of unhappiness here and hereafter. This list, he carefully committed to memory and fired it wholesale at the first delinquent waiter, who was almost paralyzed on the spot.

By this time, not a few of us were suffering from an eruption that covered our whole bodies, due to the heat. It closely resembled hives but was even more irritating.

During the afternoon of the eighth, we had a view of Monte Alégre to the north and not far distant. At 4:30 p.m. on the ninth, we passed out of the muddy Amazon into the clear water of the Tapajos, close to the little town of

Santarem. We only remained long enough to take on a supply of drinking water, but in that time, two Americans came out in a canoe.

They were remnants of a large colony that had emigrated from our Southern States at the close of the Civil War. Some of this colony were men of respectability at home, but the majority were worthless toughs who had neither the industry nor the intelligence to properly support themselves anywhere. The colony proved a poor investment to the government of Brazil, which had offered large inducements to secure them. These two men, when they came on board the *Arary*, gave a sad account of the state of destitution existing among the few remaining American inhabitants. They represented that, although they had been in the country about thirteen years, they were entirely unable to speak the language of the natives and were in a starving condition.

One of these men was permitted to accompany us and was ever afterward known by the name of Santarem.

Figuring distances on the South American rivers at the time of which we write was simply guesswork. The latitude and longitude of prominent points were approximately known, but the intervening courses and distances were absolutely undetermined. The result was a great difference in estimated distances, as given by various authorities. The distance of Santarem from the sea was thus stated by previous travelers: Herndon, 650 miles; Orton, 500 miles; Bates, 400 miles; Keller, 400 miles. The probability is that Herndon's estimate was the best.

The first case of malarial fever among the passengers occurred after leaving Santarem but was due to extreme carelessness and imprudence. There was nothing to break the monotony of our long voyage except an occasional native hut or a distant view of a troop of monkeys scampering through the treetops. The everlasting bank of green foliage that bordered the river and shut out all view of the country beyond was broken only at rare intervals by small clearings, where plantains, yuccas, and cacao were cultivated in barely sufficient quantity to supply the limited wants of some adventurous settler. We were forcibly reminded of the information given Herndon by a native in regard to a proposed stopping place, that *"Hay platinos, hay yuccas, hay todo"* (there were plantains, there were yuccas, there was everything). Add a little cigarette tobacco and a hammock, and this list would comprise all the articles the average native deemed essential to perfect human happiness.

From 5:00 p.m. on the ninth until 3:34 a.m. on the tenth, we lay at Obidos, taking on wood and provisions. Situated on a high bluff on the north bank of the river, Obidos was generally regarded as the most important town between the mouth of the Madeira and Pará. To us, it did not seem to display the business activity of Breves.

At this place, the width of the Amazon is contracted to less than a mile, and the passage was commanded by an antiquated fortification situated on a considerable elevation near the river and a short distance above the town.

Besides the usual collection of huts and hovels, the town contained some fairly good stores, an ancient church, and a school. It derived much of its importance from being a supply station and distributing point for the largest steamers engaged in the river trade. Extensive cacao plantations nearby, the fact of being a frontier military post, and its high and healthy location contributed still further to its prosperity. The many and very different species of river craft to be seen in the vicinity were alone sufficient to prove extensive commercial intercourse with remote points on the Amazon and its tributaries far beyond the confines of Brazil.

Four or five miles above Obidos, the Trombetas River enters the Amazon from the north through two mouths. This was the place where Orellana claimed to have fought the Amazons. Even in our time, the upper part of the Trombetas was little known.

Subsequently, the writer had the good fortune to meet a Jew trader, one of a class that is ever willing to sacrifice the comforts of civilization and brave the perils of a deadly climate and yet more dangerous savages in an effort to accumulate a few paltry dollars. This man had attempted to penetrate the wilds on the upper Trombetas and engage in trade with the natives. He could speak a little English and described the natives as primitive savages with painted faces who wore feathers thrust through their ears and nostrils. He summed up the whole character of the country and its inhabitants by raising both hands above his head, as though calling heaven to witness the truth of his story, and exclaiming, *"God never looks down upon that country, only the devil!"*

At 5:00 p.m. on the eleventh, we passed the village of Serpa and three hours later entered the Madeira River. Assuming the mouth of the Madeira to be 1,022 statute miles from the sea or 907 from Pará, as stated by Commander Selfridge, USN, and our average speed had been 131 miles per day, including long stops for wood, water, and provisions.

We stopped at Borba that night too late to see more than one moderately comfortable house, a dilapidated church, and about a dozen miserable huts. Probably there was nothing more to see. Borba tobacco at one time had a high reputation and was considered the best in Brazil, but in our day, it was regarded as inferior to that of Bolivia. The settlement, originally a mission, was said to have suffered greatly in previous years from attacks of the Arara Indians.

Up to midnight on the evening of the fifteenth, we made nine stops, aggregating twenty-four hours and forty minutes in the preceding four days. It is fair to presume, therefore, that on an average, at least one-fourth of our time was occupied in taking on wood and supplies.

There was little to interest anyone in the hasty trip we were making through an almost unbroken wilderness, and the writer confesses his inability to distinguish, from the steamer's deck, the various kinds of vegetation, so well described by others who traveled by canoe and much of the time close to shore.

During the entire trip up the Madeira, we did not see a single settlement that would be regarded as worthy of mention on one of our country maps at home. One of the most important was Sapucaia-Oroca, which we reached at 1:00 p.m. on the thirteenth. Accustomed to finding only large towns and places of great importance shown on our small-scale maps, one would naturally expect Sapucaia-Oroca to be, at least, a thriving and populous village. As a matter of fact, it proved to be a small settlement of Muras, Indians universally despised and notorious for thievishness and laziness. *"Lazy like a Mura"* was a common saying all along the Amazon and Madeira. The town consisted of a group of about twenty miserable huts situated on a high bank of red clay on the eastern side of the river. There, we saw them preparing parrots and toucans for the pot. A pet anteater was fastened to a stake driven in the ground. Live monkeys and parrots were abundant, and, as a verification of the principle that dogs and poverty always go together, we found a large assortment of mongrel, half-starved, cowardly curs, with scarcely energy enough to bark at a stranger. The parrots and macaws, some of them beautiful, were on terms of intimacy and familiarity with all the other domestic animals.

The monkeys and dogs played together, and in one place, we saw a parrot perched on the back of a chicken and quietly relieving it of vermin. The writer had the curiosity to enter one of the huts and was astonished to see nothing in the way of firearms. Everything seemed to indicate a degree of civilization far more primitive than that attained by the wildest tribes of North American Indians. Not a piece of metal of any size, shape, or kind was at first visible. Not a knife or spoon, not a tin dish or cooking utensil was in sight. In one corner was a bow about eight feet long, with arrows of proportionate length. In another was a *gravitana*, or blowgun for shooting poisoned arrows. Every dish was either a gourd or made of burned clay. Turning to another corner, which had escaped our notice at first, we were confronted with a *Singer sewing-machine*. These Indians were classed as civilized in contrast to the *Indios bravos*, or wild savages, who lived in the interior and frequented the depths of the primeval forest.

Some idea of the immense volume of water carried by the Amazon to the sea may be obtained when we consider that the high-water discharge of the Madeira has no perceptible effect upon the stage of water in the Amazon. Yet Keller has estimated the flood discharge of the Madeira at Sapucaia-Oroca at 1,212,286 cubic feet per second, which is nearly equal to the greatest recorded discharge of the Mississippi at Fulton, Tennessee.

The only wild animals we saw in going up the river were monkeys, occasional red squirrels, toucans, macaws, parrots, parakeets, and a bird resembling our oriole. Lizards a foot or more in length were abundant in cleared ground, where they came to sun themselves, and in one place, the writer had an exciting but unsuccessful chase after some iguanas with one foot of body and two feet of tail. They were green and as swift in their movements on trees as on the ground. From Pará to San Antonio, we saw but one single snake.

Bananas and other fruits seemed to increase in size but deteriorated in flavor as we ascended the Amazon and Madeira. Bananas, limes, lemons, and oranges were nearly always obtainable at settlements.

The oranges, though large and fine in appearance, were thick-skinned, tough, and tasteless. Chickens and eggs could generally be purchased at stopping places but only in very small numbers.

Many of us tried to catch fish, but little or no success attended our efforts, though the river was undoubtedly full of them. Dolphins could be seen frequently coming to the surface to blow, and at nearly every hut, we found jaguar or leopard skins, giving evidence of the pleasures of life in the forest.

In passing, we saw many indications that the rubber trade constituted the principal business on the Madeira, though at this season of the year, when the rubber forests, always on low ground, were submerged, there was no opportunity to witness the method of collecting and preparing this valuable product for the market.

While ascending the Madeira, all the men had their places assigned them in their respective corps. On the seventeenth, about 7:00 a.m., we reached Jumas. This was the thriftiest and most enterprising place we had seen on the Madeira. It was a Bolivian settlement and was said to have a population of 270, of whom 180 were males. They apparently spoke a dialect that was neither Spanish nor Portuguese, and few of us could understand anything the natives said.

The ground was extensively cleared, and sugarcane, bananas, plantains, and yuccas were under cultivation. We saw large quantities of Brazil nuts and a cereal resembling our rice. There was a distillery for the manufacture of *cachaca*, a species of rum as clear as distilled water and as strong as pure alcohol. It was said to be a healthy beverage but, undiluted, required cast-iron digestive organs to withstand its effects. The distillery had one still and a set of rolls for pressing the juice from the cane. Everything about the place had the appearance of neatness and cleanliness. The palm-thatched cottages were models of their kind.

We passed Crato without stopping and reached Humaita at 10:00 p.m. We were still there at 1:00 a.m. on the eighteenth, when the writer retired. There, we took on Don Ignacio Arauz and his nephew Don Santos Mercado. The former had always taken a deep interest in the improvement of the Madeira and the Mamoré and had actively assisted both Colonel Church and Franz Keller.

Our nickel-plated Smith and Wesson revolvers and light rubber coats and hats were a great attraction to the natives, and we could have sold great numbers of these articles at double or treble their value in the United States.

Before daylight on the eighteenth, we tied up at Paraiso and remained there until 10:35 a.m. This apparently thrifty settlement was situated on the east bank of the river and had a mixed population of Portuguese and Spanish Jews, with a large body of Bolivian Indians. The last were physically the finest specimens we had seen anywhere. They had large heads, intelligent faces, broad shoulders, and olive-brown complexions and were neatly dressed. The men wore white pantaloons and outside of these a short white shirt, extending a few inches below the waistline. The women in some cases were quite attractive and wore a single white muslin gown reaching below the knee, with occasionally a light calico wrapper over it and generally a string of beads about the neck. The Indians, both male and female, were barefoot. Here we saw a carpenter's shop and men making boards with a pit-saw. In loading the wood, it took seven men to carry one of our planks ashore, and in the line of about forty Indians, ten stood on this single plank.

About noon on the nineteenth, we passed the mouth of the Jamary River, which empties into the Madeira from the east. It had the appearance of a long, straight canal, and its perfectly clear water made a striking contrast with the muddy Madeira. This river had long enjoyed an unenviable reputation, both on account of prevalent fevers and the cannibal savages that lived upon its banks farther inland.

Many years previously, the Portuguese had established a

penal colony at the mouth of the Jamary and probably found it an economical institution, as two years was said to be the average life of a convict in that locality.

Late in the afternoon of February 19, high hills began to appear on both sides of the river, and at 5:00 p.m., after passing a bend in the stream, we came in sight of our destination, about three miles above.

A few minutes later, we were tied up broadside to the bank at San Antonio, six days in advance of the time fixed by contract for beginning work and having already lost our chief engineer, three valuable men, and the steam launch.

X. San Antonio

Yet I know, there lie, all lonely,
Still to feed thought's loftiest mood,
Countless glens, undesecrated
Many an awful solitude.

—Shliabhair

San Antonio, situated on the east bank of the Madeira, at a distance from its mouth of 661 statute miles, as determined by Commander Selfridge, USN, was originally a mission founded by the Jesuits in 1737 but soon abandoned on account of the fevers prevalent there. Later, there was a frontier military post for a short time maintained on the opposite, or west, side of the river, but the same conditions that made the place undesirable as a mission site rendered it untenable by the soldiers of Brazil.

Abandoned by both church and state, the pious natives were reluctant to concede greater endurance to the Prince of Darkness, and at the time of our arrival, a proverb was current to the effect that *"San Antonio was the place where the Devil left his boots."* This seemed to imply that His Satanic Majesty had also obtained an unpleasant impression of the place and had, in consequence, made a hurried departure.

Those of us who had read the accounts of previous explorers did not entertain any great expectations in regard to this northern terminus of our proposed railway, but, in view of the fact that the engineers of the Public Works Construction Company had claimed to have expended half a million dollars in the vicinity, we were naturally led to suppose that a prosperous little settlement would be found at the head of navigation on one of the greatest rivers of the world.

As a matter of fact, no one of the other sixteen places we had stopped at on the Madeira presented fewer attractions. So overgrown was the whole place with rank tropical vegetation that few of the houses in existence were visible from the steamboat and we could not even go ashore until a path had been cut up the bank, which there attained a height of about fifteen feet above extreme high water. To anyone sighing "for a lodge in some vast wilderness, some boundless contiguity of shade," San Antonio, it seemed to us that night, would prove an ideal spot. The constant roar of the rapids nearby, as the waters of the mighty river, divided by a large island, forced their way through the disrupted fragments of the immense dam of solid granite, by which Nature had once obstructed their progress, the tangled and matted mass of vines and trees that met the eye in every direction, the entire absence of cleared or cultivated ground, the two primitive palm-thatched cottages in sight, the humid atmosphere, and the sun sinking behind dark and threatening clouds, all combined to produce a gloomy impression on our minds and a realization of the serious character of the project we had undertaken to execute.

The only visible buildings were four in number. These consisted of two small warehouses, made of corrugated iron, which had been erected by the English close to the steamboat landing, and two small dwellings nearby. One of these was the store and residence of Señor Brigido (the most important personage in the place), one-story high and built in the rude native style: the framework tied together by vines, the floors and sides made of split palms, and the roof thatched with leaves of another species of palm, very abundant everywhere—not a nail in the entire structure.

The other, a little farther away, was a similar building, occupied by the *commandante* of the military forces of Brazil stationed at this point, and over it proudly floated the flag of the empire.

The *commandante* held a rank equivalent to that of sergeant, but impressed by the fuss and feathers he exhibited, most of us at first supposed him to be at least a major-general. His military following consisted of exactly six soldiers, who occupied an equal number of dirty hovels that were entirely hidden from view by the dense vegetation.

These men were occupied most of the time with their own affairs, but on important occasions, as, for instance, when the *commandante* felt it necessary to take a little exercise, they appeared properly armed and equipped to follow him, as an escort, at a respectful distance, while he paraded in full uniform, sword in hand, along the only existing path through the high weeds and brush.

Exploration and a free use of the *macheta* revealed the

existence of three other buildings, originally erected by natives for the use of our English predecessors. One of these, near the riverbank, in the middle of an overgrown and abandoned little banana patch, had once been the office of Mr. Mathews, the former resident engineer. This was in such a decayed condition as to be unfit for further occupation. The other two were on high ground about one thousand feet back of the steamboat landing, and it required a whole day's hard work, by four natives, to cut a path by which to reach them. The smaller of these two buildings was assigned to the axmen and laborers who had accompanied us on the *Arary*. The other, a two-story shed built in the native style but with two enclosed rooms, one on each floor, was adopted as temporary headquarters for the force of engineers and draftsmen to be maintained at San Antonio.

On the night of the nineteenth, we all remained on the *Arary*, but at 6:00 a.m. the next day, after a cup of coffee, the whole force was busily engaged. Supplies and baggage were being rapidly discharged. A large force of men were employed in overhauling and taking an inventory of stores left by the English in the two storehouses, of which ants, lizards, bats, and tarantulas had been in undisturbed possession for several years. Men were engaged in clearing paths and ground for camps, making the two buildings we proposed to occupy habitable, erecting tents, and conveying baggage and stores to the places on shore designed to receive them.

The engineer corps of Messrs. Runk, Byers, and Stiles went at once into camp, and the remainder of the engineers occupied the building adopted as headquarters. The laborers and mechanics established other camps, one of them known as "Edison's Camp," another as "Gorman's Varieties." It will be noticed here that, although no formal promotions had been made, Mr. Byers was in charge of a corps additional to the three originally contemplated in Philadelphia, and Mr. d'Invilliers was acting as principal assistant engineer, in charge of the headquarters corps, as a substitute for Mr. Buchholz, who was then acting as chief engineer.

That night, after supper, we abandoned the *Arary*, and, as the writer said good-bye to the officers of the steamboat and others on board, an old custom-house officer, who had been detailed to accompany us, in a burst of enthusiasm, threw both arms about his neck and exclaimed, "When the English came here, they did nothing but smoke and drink for two days, but Americans work like the devil." The immense number of empty bottles to be seen everywhere bore silent testimony to the truth of the assertion.

By the twenty-second, everything was in such an orderly condition that the *Arary* started on her return trip to Pará, and we felt justified in observing Washington's birthday by taking a much-needed rest.

Nearly every timepiece in the party had run down once or more during the trip, and as the ship and steamboat clocks were changed every day to indicate local time, no two watches set by these clocks agreed. Great confusion, therefore, existed in regulating hours of labor for the different parties. It was discovered by the Nautical Almanac, however, that on February 26, the south declination of the sun would be equal to the latitude of San Antonio and that, in consequence, the sun at apparent noon would be in the zenith. This afforded a simple, easy, and sufficiently accurate method of making our first time observation by merely noting the instant when a long plumb-line with the bob swinging in a pail of water cast no shadow.

On the twenty-seventh, Corps No. 1, under Mr. C. S. d'Invilliers, began the survey of the line from San Antonio south. About the same time, Corps No. 4, under Captain Stiles, which, until the arrival of the *Mercedita*, was insufficiently equipped for work up the river, began surveys to connect the railroad line with what was supposed to be a more desirable steamboat landing three miles below San Antonio.

On the twenty-eighth, Corps No. 2, in the charge of Mr. Joseph Byers, left in canoes for Macacos, a few miles above San Antonio, with instructions to survey the line in a northerly direction, until a junction was effected with Corps No. 1, running south.

This day, the *Arary* unexpectedly returned. She had encountered the *Mercedita* hard aground about fifty miles above the mouth of the Madeira, and Colonel Jameson had transferred all the remaining passengers and entire cargo, except railroad iron, to the river steamer. There was great rejoicing to find that Chief Engineer Bird was on board.

It appears that the launch had encountered stormy weather on Marajo Bay, and in the labyrinth of intricate channels connecting the Pará River with the Amazon, the crew had not followed the usual route, so that both the *Arary* and the *Mercedita* passed before they were aware of it. By the time Mr. Bird and his men ascertained from natives that the two steamers were ahead of them, their supply of coal, provisions, and money was completely exhausted. They then headed directly for Breves, the nearest town on the river above. On the way, several grate bars of the launch were burned out and the barrels of a rifle and shotgun, worth $175, had to serve as rather expensive substitutes.

The party subsisted mainly upon such game as could be killed with their revolvers and were compelled to cut, with hatchets, all the hard wood used as fuel on the launch.

At Breves, Mr. Bird deposited a valuable watch with Samuel Israel, an American Jew, as security for a loan of sixty dollars. He took forty dollars of this to pay his own passage on a steamer bound for the Upper Amazon, expecting to overtake the *Mercedita* before she entered the Madeira. The remaining twenty dollars he gave to his men with instructions to hail the first passing steamer bound for San Antonio and arrange to have the launch towed to that place. The accidental delay, caused by the *Mercedita* running aground in the mouth of the Madeira alone made it possible for Mr. Bird to overtake her and reach San Antonio without further mishap.

On March 1, Corps No. 3, under Mr. John Runk, was sent to Macacos and instructed to carry the surveys southward from the initial point of Mr. Byers's line.

The *Arary* again departed on the fourth and returned on the seventh with the *Mercedita* in tow and Colonel Jameson on board. The *Arary* made her third and successful attempt to return on the ninth.

The two corps under Messrs. d'Invilliers and Byers effected a junction on March 3, completing the first experimental line between San Antonio and Macacos, a distance of 4.9 miles.

On the following day, the final location of the railway was commenced and a small force of laborers was put to work grading the road bed at San Antonio.

Between the sixth and fifteenth of March, Messrs. Bird and Buchholz went twice on exploring tours up the river and reached a point above the rapids at Morrinhos. About seven miles below these rapids and some for miles above the ruins of the old English station, known as Rosstown, they established a permanent supply station and called it San Carlos.

In regard to the last of these trips, Mr. Buchholz had this to say.

We returned to San Antonio on the 15th, convinced that we must look for a cheap and good line away from the river, and that it was of the greatest importance to have preliminary surveys made as far up the river as possible, in the shortest possible time.

> All our difficulties became apparent from the experience of this trip. We had no canoes, we had no Indians—both absolutely indispensable to navigate the Madeira among the rapids and cataracts. In the absence of all roads and on account of the impossibility of penetrating the forest without Herculean labor, the river remains the only highway for transporting men, material and provisions. Our work on the river, so far, had been done by Señor Arauz, in his canoes and by his Indians, but without any definite bargain for the future, and we were at any time liable, should the Bolivian so will it, to be left in the lurch and absolutely unable to move a single corps from San Antonio. I urged at once upon Mr. Bird the great necessity of buying canoes and engaging, for permanent service, at least

fifty Indians, to be attached to the different parties in the field and assist in the transportation of supplies—a service for which these Indians are preeminently fitted and in which they cannot be excelled. Mr. Bird assured me that he was well aware of the importance of my suggestion, and had frequently spoken to Mr. Jameson about it. For some reason unknown to me, no steps were ever taken [Mr. Buchholz left San Antonio for the United States on August 23, 1878] to supply either a sufficient number of good canoes or Indians, as we were obliged to continue work the best way we could, trusting that the personal interest Arauz had in the rapid construction of the road would keep him straight and induce him to aid us all in his power. Besides these difficulties, we discovered that the provisions we had were not packed as they should have been, and as a very little forethought would have suggested.

To carry barrels of beef, pork and flour, weighing from 200 to 300 pounds apiece, tierces of hams and barrels of syrup and vinegar, weighing from 400 to 500 pounds each, over hills, while passing the falls and rapids; to lug these provisions so packed, without smaller cans or vessels of any kind to repack them, for miles over a narrow, rough path, over hill and dale to the engineer's camp in the interior, was a labor for which no provision had been made, and could not be done by any white men living; yes, in the end, wore out even the hardy Bolivian Indian, accustomed as he was all his life to carry heavy burdens.

Notwithstanding these discouraging facts, the progress of the work was far from unsatisfactory. A temporary wharf was in course of erection. Streets were being systematically laid out in San Antonio. The entire area embraced in the plan of the new town was being cleared of vegetation. The old sawmill, left by the English, was being repaired and placed in working order. Timber was being cut and hauled for the construction of new buildings then being planned in the office. In short, there was but one drawback to the general feeling of hopefulness with which we all regarded the enterprise.

We were running short of provisions and that, too, in a country where there was no visible source of supply. In a river teeming with fish, one could sit on the bank all day and cast his line without ever getting a bite. Turtles, obtainable in large numbers during the dry season when the sandbars were exposed, could not then be had at all. There was some large game in the forests, but the region wild animals had to roam over was so vast and the vegetation so dense and impenetrable that it was only occasionally that any could be seen. Even when a tapir was killed, it had to be eaten at once to keep the meat from spoiling. There were no cattle or pigs within six hundred miles. Even such things as plantains, bananas, limes, eggs, and chickens could only be had in limited quantity by sending for them in canoes a distance of from one hundred to two hundred miles.

Flour and some other essential articles were already exhausted, and we were living on preparations of moldy cornmeal originally shipped to feed cattle. Occasional cases of sickness began to appear. Colonel Jameson, though one of the last to arrive, was one of the first to suffer from an attack of the fever. Our stock of medical supplies, originally small, was daily diminishing. There was, as yet, no hospital in existence, and the sickly season was rapidly approaching. It was, therefore, with great anxiety we awaited news of the vessel that was expected to follow the *Mercedita*, about a month later, with Mr. Thomas Collins on board and additional men and supplies.

On the eighteenth, the little mail steamer *Andiri* arrived, bringing the steam launch, the three men left in charge of her, and a laborer who had been left behind at Barbados. Our consternation may be imagined when informed by these men that letters had been received at Pará, stating that the steamer *Metropolis*, bound from Philadelphia to San Antonio, had been wrecked off Currituck Beach, North Carolina, and only about fifty, out of a large number of passengers, saved.

This news cast a deep gloom over our whole community. Many of us had expected friends on that vessel. It was uncertain whether Mr. Collins was on board and, if so, whether he had been saved.

The effect of his possible loss on the enterprise itself was problematic, and our sorrow for the victims of the disaster was mingled with speculations as to how we could continue to exist in San Antonio for much longer. Many of the laborers had families at home dependent upon them for support, and now, two months and a half after leaving Philadelphia, there was no immediate prospect of receiving any pay.

A rumor became current that large bills incurred at Pará for transportation and supplies were awaiting Mr. Collins's arrival for payment. It seemed, therefore, doubtful whether further orders for supplies sent to that point would be honored, even if we did not starve before they could reach us. Altogether, the prospect was as dark as could well be imagined and calculated to check the ardor of the most enthusiastic.

There was, however, manifested by all, except a very few, a stern determination to make the enterprise a success if personal sacrifice could accomplish it. And it can be truly said that neither then nor during an even worse condition of affairs that existed later did more than a mere handful of Americans abandon the enterprise for pecuniary or any other reasons than absolute physical incapacity to continue to perform the duties assigned them.

Leaving San Antonio for a time, we shall now return to Philadelphia and endeavor, in the following pages, to trace the history of events in that locality since the departure of the *Mercedita*.

XI. The Wreck of the *Metropolis*

They prepared
A rotten carcass of a boat:
the very rats
Instinctively had quit it; there they hoist us,
To the winds whose pity, sighing back again,
Did us but loving wrong.

—Shakespeare

With characteristic energy, Messrs. P. and T. Collins had taken measures, even before the sailing of the *Mercedita*, to supplement the force on that vessel with additional men, material, and supplies, it being their intention to concentrate on the work, as speedily as possible, no less than one thousand Americans, thoroughly equipped with all necessary appliances, medical stores, and provisions.

The schooner, *James W. Wilson*, as previously noted, had been dispatched to Pará with coal in advance of the *Mercedita*.

On December 29, 1877, the steamer *Metropolis*, owned by Lunt Bros. and Co., of New York, had been chartered through H. L. Gregg and Co., ship brokers of Philadelphia. The latter firm had been highly recommended to the Messrs. Collins by Mr. Franklin B. Gowen. On January 21, 1878, the charter for the schooner *Eva I. Smith* was signed; on January 23, another for the steamer *City of Richmond*; and on February 12, still another for the schooner *John S. Wood*.

One can readily imagine that the engagement of capable men and the inspection and purchase of the cargoes for these vessels made the office of the contractors in Walnut Place, Philadelphia, a scene of unceasing bustle and activity.

On January 22, the *Metropolis* reached Philadelphia and began taking on cargo the next day at Willow Street Wharf. By the afternoon of the twenty-eighth, she was ready to start on her voyage. Mr. Thomas Collins, his wife, and several prominent employees, who had intended to leave on the *Metropolis*, at the last minute, decided to wait for the *City of Richmond*, because the superior speed of the latter vessel would enable them to reach San Antonio quite as soon as by the former.

The *Philadelphia Times* of January 29, 1878, contains the following account of the sailing of the *Metropolis*.

> The steamship *Metropolis*, Captain Ankers, the second dispatched by Messrs. P. & T. Collins to Brazil, sailed yesterday with 215 passengers, 500 tons of rails and machinery and 200 tons of stores, all in charge of Mr. Paul J. White, late chief engineer of the Lehigh Navigation Company, and Mr. James T. Moore, also an engineer of reputation and experience. The passengers were principally laborers and foremen engaged to work on the Madeira and the Mamoré Railway. ... The long wharf of the Reading Railroad at Willow Street, at

the end of which the departing steamer lay, was filled with an immense concourse of people, whose numbers could only be computed by thousands. The majority of them were Irish; men, women and children, assembled to give a long and sad good bye to the husbands, fathers and friends, who were about to bury themselves, for eighteen months at least, in the dense and somber forests of the Upper Amazon, and the scenes were such as the wharves of Queenstown and Londonderry witnessed in the days of half a century ago, when the stalwart sons of the Green Isle, bade tender and tearful farewell to their friends and relatives, as they embarked on the vessels which were then about to convey them to a land as much a *terra incognita* to them at that day, as the heart of the South American Continent is to the emigrants of yesterday. The incidents as the passengers embarked were frequently pathetic, the sorrowful parting of wives and children bringing tears to the eyes of bystanders long unused to the melting mood. One fine looking woman, the wife of one of the foremen, after a dozen passionate farewells, finally clung to her husband with such intense sorrow that the latter was compelled to remain on the wharf.

A stalwart Irishman clasped to his heart a fine manly boy of seven years and begged with tears in his eyes for permission to take him along. "You won't part us," the father entreated. "It is impossible for you to take the boy," replied Mr. Collins. "Then I can't go, I won't go," said the father. The little fellow knew of the stern want that was driving his father away to a distant land to earn the bread so hard to win here and the future man welled up in him, as he said, in spite of the sorrow of parting, "Pap, go, an' I'll stay with Uncle Jim; he'll take care of me," and the father, then by far the weaker of the two, sailed away. There were on the wharf and on the steamer embracings and kissings and tears and cries of sorrow until the steamer cast off her moorings and started on her way to the sea, and such a pressure at her gangway that several of the passengers were left behind and personates taken away in their places.

The *Metropolis* sailed out into midstream, and there anchored until the morning of the twenty-ninth, when she steamed down the Delaware. She reached the breakwater at about 11:00 p.m. the same day, and there dropped her pilot and Mr. Joseph A. Connelly, a clerk for P. and T. Collins, who had gone that far to perfect the roll of passengers and assist in the formation of messes. Mr. Connelly reported no indication of bad weather at the breakwater and that the passengers were all in good spirits. From that time, nothing was heard of the vessel for several days.

On February 1, the *Philadelphia Times* published the following telegram:

Norfolk, Va., Jany. 31, 1878.

At 6.30 this afternoon[10] the steamship *Metropolis*, from Philadelphia for Pará, Brazil, went ashore on Currituck Beach, three miles south of the lighthouse, during the prevalence of a furious southeast gale. Great confusion

---

[10] The wreck occurred in the morning, about 6:45, but the news was not sent until long after.

prevailed on board. Owing to the fury of the gale and the roaring of the surf the orders of the officers could not be heard. About fifty of the passengers and crew were washed ashore. About two hundred are believed to be lost. From some of the sailors, who arrived at one of the signal stations, it appears that the vessel had encountered heavy gales from the southeast for the last twenty-four hours and when she struck she was heading about south-southeast. The vessel swung broadside to the surf, which made a complete break over her and washed many of the people overboard into the sea. Reporters have gone to the scene, *via* the Albemarle and Chesapeake Canal by the tug *Crotian*.

The receipt of this news caused the most intense excitement in Philadelphia. The relatives of the passengers nearly all lived in the vicinity, and during the period of anxiety and suspense that followed, they gathered in front of P. and T. Collins imploring piteously for information in regard to the fate of individual passengers, which, for a long time, could not be given. The senior member of the firm happened to be absent from the city at the time the news of the disaster arrived and upon Mr. Thomas Collins devolved the painful duty of responding to the personal and telegraphic inquiries of grief-stricken relatives, giving information to reporters, receiving and replying to messages from the telegraph office nearest to the wreck, and making all arrangements for the relief of the survivors, as well as caring for the bodies of the dead.

He was deeply distressed by the manifestations of anxiety

and grief he was compelled to witness for several days, and, to add to his trials, one or two newspapers intimated that his firm were in some way responsible for the disaster, an intimation that no facts, then known or afterward discovered, could, in the slightest degree, justify. Notwithstanding the extreme difficulty of reaching or communicating with the scene of the wreck, the *New York Herald* correspondent telegraphed from Norfolk, Virginia, under date of February 1, an account of the disaster that had occurred on the preceding day. For length, accuracy, and minuteness of detail, these dispatches constituted a feat in journalism that, so far as the writer is aware, has never been equaled at any time or in any part of the globe.

The results of subsequent investigations did not in any important particular necessitate a modification of the statements they contained. The extracts here given are only a small part of the information regarding the wreck contained in the *Herald* of February 2, 1878.

The Correspondent's Story

Norfolk, Va., Feb. 1, 1878

So far the news from the wreck of the steamer *Metropolis* is exceedingly meager. Only one survivor has arrived in this city and he, being a bricklayer, is able to tell little beyond his individual experience in escaping from the vessel.

Following so quickly upon the *Huron* disaster, the *Metropolis* affair has created a profound sensation here, as the vessel was widely known to be unseaworthy. It is appalling to realize such a terrible loss of life and the craving for even the slightest evidence is both startling and agonizing.

The scene of the wreck is about sixty or seventy miles from Norfolk and is almost inaccessible either by land or water. It is much worse in this respect than was the *Huron* wreck, which could be reached through the canal and Sound much more easily. But the *Metropolis* cannot be reached except by tugs drawing, at the most, two feet and a half of water, or by a tedious journey over a bad road, a rough and uneven beach and every other obstacle known to travelers. The history of the sailing of the *Metropolis* from Philadelphia and the mission in which she was engaged are already familiar to the readers of the *Herald*. She was heavily laden with railroad iron and also a living freight of 200 persons, including three females.[11]

After discharging her pilot at the Delaware Breakwater, the weather became heavy. A strong northeasterly gale sprang up, which tried the ship severely and, in the very beginning of her voyage, both passengers and crew began to entertain serious misgivings, as to the seaworthiness of the vessel. On Wednesday night she sprang a leak. This created a panic, which was in part subdued by a vigorous use of the pumps. All hands went to work at them with that energy which is only imparted by despair, but despite their efforts, the water increased in the hold. The pumps

---

[11] The investigation afterward proved that the vessel carried a crew of twenty-four men, passengers to the number pf 221, and one stowaway, 246 in all (NBC).

failing to work, Captain Ankers proceeded to lighten the cargo by throwing overboard all the heavy material possible. Still, the water continued to rise and death began to stare the panic stricken passengers in the face. The leak increased and the pumps proved powerless to keep the Arara Indian's vessel clear. Captain Ankers then headed the course of the ship to that haven of refuge for disabled vessels, Hampton Roads, but, as fate would have it, he mistook his reckoning. This was owing principally to the heavy and blinding snow storm that prevailed at the time and the back water current of the Gulf Stream, which set the ship to the southward and westward.

The fires were then nearly out, but the vessel still bore on until it became a terrible necessity with the captain to run her ashore, with the hope of at least saving some of the precious lives of the passengers and crew. At last the fearful and much-wished-for moment arrived and the *Metropolis* struck the shore in the vicinity of Whale's Back Light, on the Currituck Beach about twenty-five miles to the north of where the *Huron* went ashore, three miles south of the above mentioned light and between Life Saving stations Nos. 4 and 5. Then ensued a terrible scene of panic, terror and confusion, such as only those who have had the experience of a shipwreck can imagine.[12]

Just before the ship beached she shipped a heavy

---

[12] Daniel M. Cozzens, second officer of the *Metropolis*, subsequently testified, "There was no excitement whatever on board the *Metropolis* during the trying hours that preceded her demolition."

sea, which carried away her engine room bulkhead and forward cabin. The sea poured into the vessel in an awful volume and extinguished her fires. She was then a helpless monster, utterly powerless to stand the seas that swept over her and beyond all human skill to manage. At about a quarter to seven o'clock on Thursday her course being about south-southwest the *Metropolis* struck. She broached broadside to the beach, which caused every succeeding sea to sweep her decks, carrying into the seething foam the unfortunate passengers, almost by the score. During all this agonizing scene of terror and suspense a man on horseback was seen in the dim light on the shore waving his hat, which gave some faint promise of assistance, at which the living passengers eagerly grasped, but, alas, no help came. Many who had held on to life, finding the expected help not at hand, gave up in despair and sank to rise no more.

The terrible moments, which seemed to the survivors like ages, wore on until about eleven o'clock, when, at last, the life-saving crew put in an appearance. But they were unable to render any assistance. They were merely idle lookers-on at the havoc of death among the wild waves, which seemed to sing a requiem for the despairing mortals, who were so rapidly drowning. The most that these inefficient Government employees did was to pull some of the more fortunate out of the surf, who had been washed ashore on pieces of the wreck.

With the striking of the ship commenced a scene which defied description. No orders could be heard, and if heard no one was willing to obey. The one feeling, that each one must save himself or drown, mastered every other. A rush was made for the life preservers, of which there was a full supply on board. A struggle ensued of the strong against the weak for the coveted articles. In the end, however, none of the preservers could be obtained, nearly all of them having been washed overboard by the tremendous seas that swept the ship before they could be used. The angry waves each moment decimated the hapless sufferers, who, for the most part entirely unused to life on shipboard, knew not where to seek protection, or to avail themselves of opportunities of safety which occurred to more experienced persons. Captain Ankers and most of his officers stood by the ship as long as possible. The survivors speak of his efforts as being almost superhuman during the time the ship was struggling with the gale, while she was disabled and drifting helplessly to certain destruction and after she had struck upon the treacherous sands.

The wreck broke up fast. Captain Ankers manfully did his duty, giving orders, helping and cheering everybody. He was about the last man to leave the ship. He hoped against hope, expecting succor from the shore. Had he received this, many precious lives would have been saved and few would have been lost.

It seems that the first mate, Mr. C. B. Dickman, was among the first to reach the shore and he immediately dispatched a messenger to Kitty Hawk Station, twenty miles to the southward, imploring speedy assistance, but it was six o'clock on Thursday evening before any

Pennsylvania's Amazon Princess Railroad

dispatch was forwarded to the Signal Service Bureau in Washington, and, in the meantime nothing could be done to assist those on board.

The only efforts that could be made were to save those, who, in attempting to swim ashore, were floundering helplessly in the surf and to drag them half drowned to dry land. The fate of Captain Ankers was for some time unknown to those on shore and it was by them supposed that he had gone down with the ship when she finally went to pieces, but he subsequently made his appearance, to their great relief, having been washed ashore several miles northward of the stranded vessel. The condition of the men who had succeeded in reaching the shore with him was deplorable in the extreme. Half naked and without fire or shelter, they suffered bitterly from the cold for several hours until they could be furnished with assistance by the kind hearted, but much maligned people in the vicinity.

The people of Norfolk, at a meeting called this evening by John S. Tucker, made every arrangement to supply the survivors with all they may need, and in so doing, have nobly sustained their reputation. A large store was secured to receive the parties upon their arrival by the steamer *Cygnet* tomorrow night, when she is expected to bring up Captain Ankers and the 151 now known to have been saved. Dr. Sawtell, of the United States Marine Hospital left this afternoon on the steamer *Haring* for Coinjack, N.C., with a supply of clothing, medicine and provisions, under orders from Washington to render all the assistance possible. The naval authorities also dispatched a steam launch for the scene of the wreck *via* the Albemarle and Chesapeake Canal in command of Captain Gillis of the receiving ship *Franklin*.

The Barker Wrecking Company have also dispatched their wrecking tug *Resolute* to the spot by the outside route. Assistant Postmaster, Samuel E. Shipp, of this city, under orders from Washington left on the steamer *Haring* for the purpose of looking after the United States mails for Brazil, if by any possibility any of the bags may have drifted ashore. In order to afford every facility the Signal Service has erected a telegraph station on Bold Beach, where its operators have remained manfully at their post day and night, without rest or relief since yesterday morning. The scene along the beach for miles north and south is said to be distressing to the last extreme. The shore is strewn with pieces of the wreck and bruised and battered bodies constantly appear, taking the place of those which have been tenderly taken up and interred. A large number of the survivors are more or less injured, some of them being entirely disabled from collision in the water with drifting pieces of the wreck. They are scattered all over the country for miles, wherever accommodation could be obtained, and the nearest to the scene of the disaster are not less than three miles away. From present indications there is little probability of anything of value being saved from the wreck. The ship has gone to pieces and the coast so beset with quick sands, that before the sea goes down sufficiently to allow working upon the wreck, all the heavy articles will have become too deeply imbedded in the sands to be recovered by any appliances in use.

The following is the statement of Richard W. Brooks, of 227 Price Street, Germantown, Pennsylvania.

Statement of Richard W. Brooks[13]

We have only three women on board, one of whom was the wife of the chief engineer, another the wife of a Mr. Harris, the steward. The third one's name was not known. I and a crew of twenty-seven men all told, left Philadelphia about 4 P.M., on Monday, January 28th, and lay in the Delaware until 9 A.M., Tuesday, January 29th. At the Breakwater we left the pilot and Mr. Connelly, the clerk to Mr. Collins. Both wished us *bon voyage*. We proceeded to sea. When some miles out from the Breakwater at 5 P.M., Wednesday, the first mate found the ship had sprung a leak and the men were immediately put to work throwing out coal to lighten her. After getting rid of coal, fifty or seventy-five tons, and being then unable to find the leak, Captain Ankers ordered all hands to put on life preservers, as the ship was making water very fast and putting the fires out. On Thursday morning, about three o'clock, all hands were called to raise sail and the ship was headed directly for the lighthouse, which was supposed to be St. Charles Light. At 7.30 A.M., about two hundred yards from the shore, the ship struck.

---

[13] Brooks was a bricklayer by trade and one of the passengers (NBC).

The waves immediately commenced washing the decks from stem to stern, breaking in the weather side. We all stood this for about two hours, when one man, a fireman, started to swim ashore, also, the first mate, after whom I followed. We managed to get ashore after a hard swim, landing about half a mile above the place where the ship had struck. Found no one on shore to afford us assistance. After we three men reached the shore all were so weak as to be scarcely able to move. A boat with six men put off from the ship and succeeded in reaching the shore with all on board. Nine of us then started up the beach, following the telegraph poles, in search of some one to render us aid. The only living thing we saw was a cow and she immediately ran away. We, however, followed her until we got on a sand hill, from which we saw the lighthouse, and, after running a mile through brush and briars in our bare feet, with nothing on but shirt and drawers, we came to the Currituck Light House Club. Mr. William Jones, one of the employees, received us and told us to take off our wet clothes and he would give us dry ones, after doing which, he immediately started a boy on horse back to the lighthouse to inform the keeper of the wreck and to send the life-saving apparatus to the scene.

After getting on dry clothes, we all proceeded to the wreck accompanied by Mr. Jones and four men belonging to the Life-Saving Station, who carried ammunition and a mortar. We followed the beach and found it covered with bodies and portions of the wreck. Upon getting opposite the ship the mortar was placed in position and a line was thrown over the topsail of the vessel. The under current was so strong, however, that it snapped the line. They again fired two more lines towards the ship, but failed to hit her. These three shots exhausted all the ammunition, and the men, finding it useless to expect any help, commenced trying to swim ashore. The fourteen men on shore then formed a chain and advanced into the water to aid our unfortunate ship mates. Of the first five who tried to swim ashore, we saved four. The men then commenced leaving the ship, one and two at a time, until twenty-six were saved by us.

About one o'clock, Thursday, the ship began breaking to pieces rapidly. At that time all that was left was the hurricane deck over the saloon with as nearly as I could judge, one hundred and fifty men holding on to the rigging. I then left the beach and went to the Club House, picking up a suit on the way which I put on to save myself from unnecessary exposure, although they were very large. I then took the boat with Mr. Jones and went over to his neighbor, Mr. Hampton, where I got my supper and stayed all night.

On Friday morning, Mr. Hampton took me in his row boat and carried me to Currituck Landing, where I took the *Cygnet* at 6 A.M. and proceeded immediately to Norfolk to communicate with Mr. Collins in regard to what I should do for the comfort and relief of the survivors and with the remains of those lost. I telegraphed Mr. Collins from the Canal Locks as follows: *"What shall I do with the men who were saved."* To this he responded, *"Provide for them. I will be on tomorrow."*

From the dispatches of the *Herald* correspondent, sent after he had visited the wreck and dated Norfolk, February 2, 1878, we make the following extracts to complete the story of the disaster.

The whole coast was covered with the ghastly evidences of this fearful wreck for miles to the northward and southward of it. It is gratifying to recount that, amid all the scenes of woe and death, all on board exhibited a heroism and coolness worthy of the highest admiration. The captain behaved with the greatest gallantry and his conduct was an example that encouraged and cheered the passengers and crew ... It is beyond the power of description to picture the misery and wretchedness of the rescued, whose pitiable plight in nakedness and utter destitution cannot be portrayed. They were huddled together in groups to keep warmth in their bodies. Numbers of them were wounded and bruised and all of them were hungry and starving ... The story of the wreck is one of horror. It speaks in no uncertain words and will cause a just and righteous indignation against those who put at the mercy of the ocean waves 250 lives in a vessel that was in every respect unseaworthy, that was as rotten as punk and that could well be likened to a death trap ... A like censure is due to the miserable apology for what is termed the United States Life Saving Station. It has on these two memorable occasions, of the *Huron* and the *Metropolis* wrecks, been proved utterly worthless ... The route over the beach was thickly studded with the newly made graves of victims. They were marked with rude boards and in some instances the bodies were buried five and six in a grave. Only in the immediate vicinity of such a disaster is it possible to realize the true import of the terrible word "wreck." In the name of humanity and civilization!

It was a ghoulishness almost incredible, committed mostly by Negro residents of the beach and close by, aided by a number of whites. They took all the valuables and even the clothing from the dead bodies that had been washed ashore. The sacred person of a female was not a bar to the worst sort of barbarism ... To-day at twelve o'clock the survivors were put on board the *Cygnet* and they will reach Norfolk to-night, where they will be taken charge of by a special committee of citizens and provided with comfortable clothing and food.

Statement of Charles B. Dickman, First Mate of the *Metropolis*.

The steamer *Metropolis* left Philadelphia on Tuesday at eight o'clock in the morning. We passed the Breakwater at 11 P.M., discharged the pilot at half past eleven and proceeded to sea. The weather was fine and everything favorable for a pleasant voyage. At one o'clock in the morning the wind commenced to blow from the eastward and, by seven o'clock, it had increased to a fresh gale, with a bad sea and ship laboring greatly. We reduced sail, and in the meantime, the gale had increased to so great an extent that we were forced to take in all sail.

At eight o'clock Wednesday evening we discovered that the ship was making water and leaking badly. We then started all the passengers, 220 in number, to pass up coal from the after hold and headed ship towards Hampton Roads. At this time the steamer was about 60 or 70 miles east from Cape Henry. The men worked well and faithfully and, by midnight Wednesday, we had discharged all the coal in the after hold and a good part of what was obtainable from the main hatch.

By this time the engine was hardly able to keep the water down, although the carpenter and his force were employed all night in trying to mitigate the leak, which had previously been discovered above the stern post. At 2 A.M. Thursday the circulating pump gave out and the engineer had to run high pressure, which, of course, stopped the power of the pumps as the water was gaining. We made the light (Cape Henry) about 8 A.M. on the 31st. The weather was very thick and we hauled off shore, head to the wind in twelve fathoms of water to await daylight.

The wind and sea rising and there being no possibility of keeping the steamer off land, as she was breaking up in her upper part, with her boats all gone but one, and finding no command could be had of the vessel through lack of power, Captain Ankers, after consulting with his officers, decided to head her for the beach, in the hope of saving some lives. At this time the storm was raging furiously and heavy seas coming on board. The forward saloon had been washed away as, also, the temporary house over the forecastle. During all this time the officers and crew discharged their duties with coolness and the passengers behaved admirably. At 4.30 A.M. an immense sea boarding the steamer carried away her smoke stack and engine room doors, and in consequence a great mass of water found its way below, partly putting out the fires. After the steamer was headed for the beach we kept her before the wind with all sails set and by 6 A.M. the fires were thoroughly put out. At 6.45 we struck the beach about three miles south of Currituck Lighthouse with head on and bow not 100 yards from the beach.

There were 246 souls on board at the time the vessel struck and, up to noon that day, there were no signs or appearance of any life-saving crew to render assistance, although they were only three miles from the wreck, Station No. 4 being just at the lighthouse. About twelve o'clock, the life-saving crew arrived at the scene of the wreck and shot two lines. The lines missed. The second was caught, but owing to the stupidity of the crew on the beach the line parted. After making two more abortive attempts to throw the line, the ammunition gave out.

During this time a number of the officers, passengers and crew had jumped overboard and swam ashore. A number of them were drowned. I was one of the first to reach land. Previous to my leaving the ship every one had been provided with a life preserver. Captain Ankers, during the whole time, displayed the greatest courage and coolness, superintending all details and acting with such prudence and judgment as caused all to endeavor to emulate his example. There were three ladies on board, who were especially provided for. During the whole time the ship was washed by immense seas, which swept away completely the after part of the hurricane deck, and the people were forced to take to the fore-rigging for greater safety. Occasionally three or four persons would jump overboard and swim ashore.

One of the ladies, Mrs. Meyers, wife of one of the foremen of laborers, was washed out of the pilot house, knocked senseless against the rail and drowned. Her husband was near her at the time. Shortly after he jumped overboard and, owing to the misappliance of the life buoy used, he was also drowned.

After the failure of the life saving crew to get lines to the ship, signals were made to those on board to try to reach the beach as best they could, arrangements having been perfected on shore, as far as could possibly be done, to save them. By this time the ship was breaking up very badly, parts of her top sides having been stove off. Numbers then commenced to leave the ship, using every facility for so doing. In rescuing these, it is only justice to say the crew of the Life Saving Station rendered all the assistance they could, but had no boat … At three o'clock in the afternoon the sea became very violent and the ship broke up rapidly. By five o'clock the fore-mast went on the board and all those who were then on the wreck, about sixty in number, were swept off. The captain and second officer were the last to leave the ship.

Captain Ankers from the first refused to leave the ship until all hopes were gone and not while a plank was left to stand on did he leave the vessel, remarking to those who had remained with him that they now must "strike for the shore." Too much credit cannot be given to the engineers of the steamer who stood by their posts to the very last, until the fires were extinguished. They acted nobly.

Nothing that could now be said would add to the thrilling tale of disaster contained in the foregoing statements of cold fact, made by trained observers on the spot and by the surviving victims who were washing ashore on that fatal January 31, 1878. Captain J. H. Ankers was laid up with an attack of pneumonia soon after the wreck and not till long after could a statement be obtained from him. Several facts of interest became known later. One was that seasickness among the passengers and the weakness resulting from it had much to do with the loss of life.

The *Metropolis* carried no cannon for signaling purposes. Had she been provided with one, she could easily have attracted the attention of the crew at the nearest life-saving station, which was only three miles distant, and enabled them to arrive at the wreck fully five hours earlier. The loss of life, as ultimately determined, was much less than early reports indicated.

The crew consisted of twenty-four men; out of that number, ten were drowned. The employees of P. and T. Collins numbered twenty-five saloon and 196 steerage passengers. There was also one stowaway, making 222 passengers in all, and of that number, as nearly as could be ascertained, from seventy-five to eighty lost their lives. The larger number was probably the correct one.

It was stated in the newspapers that Mr. Edward Lafourcade of Philadelphia, who went out as a clerk for the contractors, was the only person on board who reached shore attired in all his usual winter clothing, including a heavy overcoat. This gentleman subsequently testified under oath that *he did not bring his valise off the vessel with him.*

It is unnecessary to describe the scenes attending the arrival of the survivors at Broad Street Station, Philadelphia, on February 5. The generous sympathy and material aid extended to them from all quarters, both on the way and after their arrival, can always be expected from the American people when their hearts are touched by reading the terrible details of some great public calamity. Indeed, it was said, that, for many days thereafter, some unworthy citizens, who usually frequented a winter resort maintained by the city on the banks of the picturesque Pennypack, enjoyed a life of luxury and ease by representing themselves as survivors of the ill-fated *Metropolis* and rehearsing for the benefit of a too-credulous people the harrowing scenes they had familiarized themselves with by diligent perusal of the newspapers.

The genuine survivors were, however, made of sterner stuff, and the New York *Herald* in February records the fact that a majority of them had applied for permission to sail on the next vessel for Brazil. At 5:00 p.m. on February 12, the steamer *Agnes* from Norfolk arrived at Philadelphia with the resurrected bodies of those who had perished. This event brought to a climax the popular indignation, already intense, against the owners of the vessel and the New York inspectors.

This feeling found expression in the newspapers from one end of the country to the other. As usual, upon such occasions, overwrought sympathy and an insatiable desire for vengeance upon those really or supposedly responsible for the destruction of human life are concurrent psychological phenomena, which emphasize too much intensity for long-continued existence. The wreck of the *Metropolis* was one among the many previous and subsequent exemplifications of this social habit. Many allegations of criminality on the part of the owners and New York inspectors were published, among which none were more pointed and specific than the following, which appeared in the *Public Ledger* in Philadelphia on February 16, 1878.

> The following taken from the documentary history of the *Metropolis* tells its own story. The *Metropolis* obtained her first register as the *Stars and Stripes*. Her first register was issued at New York, May 22, 1861, as having been built at Mystic, Connecticut. Her length was then 147 feet, beam 34 feet, depth 9 feet, tonnage 407 tons. She was sold to the United States in July,

1861. She was purchased from the United States in September, 1865, when no evidence was produced of the time or place of building. She was remeasured and got her next register from the Secretary of the Treasury September 5, 1865. Her length was then recorded 142.9 feet, beam 35 feet, depth 16 feet, tonnage 484 tons, new measurement. There were several changes of papers after this. In May, 1871, when an enrollment was issued at New York and surrendered July 20, 1871. The cause of surrender was "Broken up."

She next appears as the *Metropolis* with temporary registry issued at Newburyport, Massachusetts, April 28, 1871—one month and eight days after surrender of her enrollment as the *Stars and Stripes* listed on account of "vessel broken up." Her length was recorded at 198.6 feet, beam 34 feet, depth 16 feet, tonnage 879 tons.

She was to have her permanent register on certificate of the master carpenter, as a pencil note on the original application of the register indicates, but that certificate was never obtained. *She got her permanent certificate on a nondescript oath of Benjamin P. Lunt and John Hageman, Jr., that the vessel was built by them in August, 1871, at Newburyport, Massachusetts.*

This statement, so far as the writer is aware, was never validated; it was supported by statements made in the New York *Herald* and other papers. It indicated a change of name, which is in itself a violation of a United States statute;

a lengthening of the old vessel to the extent of 55.7 feet; an increase in her tonnage of nearly ninety-five tons; and a certificate obtained by representing the vessel to be new, when in fact, she was seventeen years old at the time of the disaster. Other allegations were made to the effect that the lengthening had been made by cutting the vessel in two, drawing the pieces apart, and inserting about fifty-six feet of new work in the middle between the rotten parts of the original structure.

Secretary of the Treasury John Sherman launched an immediate and searching investigation, which was conducted at Norfolk, Philadelphia, and New York. The testimony elicited was so conflicting and contradictory that it can with difficulty be explained on any other hypothesis than that of unblushing and shameless perjury by one side or the other. It is only fair to say that nearly all the expert testimony went to prove the seaworthiness of the vessel. The owners, the New York inspectors, the ship's officers, the men who had insured her, and, in a qualified way, the men who had insured the cargo all asserted under oath her fitness for the voyage.

The most positive evidence was given, on the other hand, by many witnesses to the rottenness of the ship. It was stated that she was "as rotten as punk." One person reported having pulled pieces out of the deck with his fingernails; one said that she broke up like kindling wood and that many pieces found on shore could be crushed into powder by hand. To all this, the answer was made that "all vessels were more or less rotten."

By some witnesses, the disaster was ascribed entirely to the rotten condition of the old parts of the vessel, which, they claimed, were the cause of the leak that forced the captain to run on the beach.

The owners endeavored to place a large measure of the blame on the government on account of the inefficiency of the life-saving service. Both parties to the investigation were, therefore, on the defensive, and no one appeared as the special representative of the unfortunate victims. The owners claimed that no vessel could have outlived such a storm, though no evidence was produced to show that other wrecks had occurred at the same time.

The New York Board of Underwriters had rated the vessel at A 1-1/2, or second class. The Philadelphia Board of Marine Underwriters had rated her as O 2-1/2, or fourth class. Placing her in the fifth class would have been equivalent to condemnation, as vessels of that class were never insured.

P. and T. Collins had taken out insurance on the cargo to the amount of $42,000, much less than its actual value. The owners had insured the vessel in New York for $18,500 and testified that they valued her at $160,000. Captain Monroe, surveyor and inspector for the Philadelphia Board of Marine Underwriters, testified, "I should judge the value of that boat would not be over $12,000; that means everything," as she was on arrival at Willow Street Wharf.

Two reputable men, Thomas J. Cogan and George A. Yohe, both in the employ of P. and T. Collins, testified that Captain Ankers had told them that his last instructions from one of the owners were: "Captain, if you meet with any accident make no half-way job of it, make a complete wreck."

On February 15, Cogan and Yohe sailed from Philadelphia for Brazil, and two days later, a letter from Captain Ankers was published in the *New York Herald* in which, after expressing his regret at being physically unable to attend the investigation and give his testimony, he flatly contradicted Cogan and Yohe, denied that he had ever received any such instructions as they stated, and, in the most positive terms, declared that the statements of these men were absolutely false.

Cogan and Yohe, of course, knew nothing of this contradiction for months afterward, but the vessel in which they sailed carried a special correspondent of the *New York World*, who wrote to that paper a long letter from St. Thomas describing the voyage and giving sketches of the prominent passengers.

This letter was dated February 23, 1878, and was published on the following March 13. In it, the correspondent mentions having met several survivors of the *Metropolis*, including Yohe and Cogan, and refers to the latter as "having a remarkably good memory." With Cogan's permission, he gives his story for the voyage and wreck of the *Metropolis*. The correspondent introduced the remark that "This crooked business needs to be so thoroughly investigated as it deserves for these gentlemen seems to me very informed."

Cogan is quoted as saying at the close of his statement, "The next day (after the wreck) I saw then that it became evident to everyone that her deck was rotten and like punk tinder ... My impression was that it was a foregone conclusion on the part of the owners that the ship would never reach her destination." Thus, he provided a reason for the opinion he had formed and reiterated the remark attributed by him, in his testimony, to Captain Ankers, which the captain later denied ever having made. The correspondent said, "I believe statements of this descriptive comment, but I cannot forbear asking, when will the world's justice be weighed on the same scales, as the petty thief and ordinary murderer, the million robber and wholesale murderer."

The *Philadelphia Times* (which was owned by the Collins Brothers) of May 6, 1878, received a telegram from Washington, which stated that a report of the supervising inspector general concerning the case of the *Metropolis* with accompanying papers had been received by the secretary of the treasury and referred to the Treasury Department, with instructions to institute such legal proceedings against the owners that the evidence in the case seemed to warrant.

This telegram concludes as follows:

> Mr. Lunt, the principal owner of the vessel has been here for several days endeavoring to have the second report reviewed and also claiming it does him great injustice and is in non-conformity with the facts. He has been invited to make his request in writing, with any corrective evidence he may have to present. This he has a disinclination to do and, in the meantime, states that notice will be taken of his oral representations.

There is nothing in the newspapers to indicate that litigations had accomplished anything. The Philadelphia insurance company addressed a letter to the secretary of the treasury in Washington, asking about the result of the proceedings instituted by his predecessor in this matter.

The inquiry was referred to the Department of Labor and then to the life-saving service and to the supervising inspector of steam vessels, from whom he received a reply, dated January 14, 1879, containing this information.

> Being in receipt, by reference, of your letter of January 14th, in the matter of the investigation of the vessel *Metropolis*, wrecked January 31, 1878, and he would say that, while there is a mass of correspondence on record in this office in regard to the case being investigation seems to have been long drawn out, very much complicated, I am unable to find a final decision in the matter, so far as the records indicate.

It is not the writer's purpose to indicate where the blame for this disaster should be placed. He records the facts, as widely published at the time, and the reader can decide for him- or herself, whether it was due to fate, beyond human control, or to the same unpunished, grasping greed that, beyond a doubt, caused the *General Slocum* holocaust and is but one more illustration of the operation of that predatory law of nature mentioned by Bayard Taylor, that:

> Each prodigal life that is wasted
>   In manly achievement unseen,
> But lengthens the days of the coward,
>   And strengthens the crafty and mean.

XII. Departure of the *City of Richmond*

To set out betimes is the main point.
—La Fontaine

The unfortunate wreck of the *Metropolis* made it imperatively necessary to hasten the departure of the next vessel of the expedition. The morning of February 14, 1878, saw the steamer *City of Richmond* at Willow Street Wharf, Philadelphia, waiting to receive her 40 crewmen and 427 steerage passengers.[14] Besides a large consignment of baggage, provisions and general supplies, she carried 620 tons of railroad iron and 235 tons of coal.

In a scan of the list of cabin passengers today, a handful of names appear that deserve passing notice. At the top of the list was the name of Mr. Thomas Collins of the contractors. Another was that of Othniel E. Nichols, since chief engineer and general manager of the Brooklyn Elevated Railroad, principle assistant engineer on the Williamsburg, Pennsylvania, bridge, and chief and consulting engineer for

---

[14] The captain intended that a larger number of steerage passengers were available, but this seems to have been the actual number carried.

the Department of Bridges, New York City. Mr. Nichols was then a comparatively young man, thirty-some years of age, physically in his prime and already guardian of an enviable reputation.

In obtaining the degree of civil engineer at Massachusetts Polytechnic Institute in 1868, he had been for a time employed on work connected with the bridges and water supply of Prospect Park, New York, as assistant engineer on the West Side Tram (old Greenwich Street) Elevated Railroad and in which latter capacity he erected the first elevated train of New York City north of Cortlandt Street. Also, he had been engineering secretary to the then president of the New Jersey-based Iron Company. In December 1871, he worked in Peru as assistant engineer on the Callao, Peru, Oroya Railway, where he remained two years as engineer of a mountain section of heavy rock and turn around.

In 1871, he was division engineer responsible for the location and tunnel work on the Huaraz Railway, a narrow gauge line in total about 160 miles in length. After returning to the United States in 1876, he was employed as the assistant engineer and superintendent by the New York Bridge Works, then holding a contract for the construction of the Sixth Avenue Elevated train of New York City. In 1877, he was assistant engineer of the Department of Parks and in the final construction of the Brook Avenue sewers for the Borough of the Bronx.

Mr. Charles F. King, at that time just married, since a leading contractor on the (US) National and other railways, later a surveyor from Schuylkill County, Pennsylvania, and being of a well-known firm of Philadelphia railroad contractors, was also one of the passengers.

Messrs. Collins and King were accompanied by their wives, and Mrs. Nichols was subsequently to join her husband at San Antonio. We shall have more about those ladies in a succeeding chapter. It is enough for the present to say that their many kindnesses to the sick as well as their uncomplaining endurance of the unavoidable hardships, incident to the tropical region far removed from the comforts of civilization, won for them a high place of regard and esteem of all connected with the enterprise.

One of the medical officers on board was Dr. Isaac T. Coates of Coatesville, Pennsylvania, who, in addition to being an eminent physician and surgeon, was possessed of no mean ability. He had already had a full, eventful career. He was a surgeon in the navy during the Civil War and of the Seventh Cavalry under General Custer's first western expedition, acting at the same time as correspondent for leading eastern newspapers. He was with Henry Meiggs four years and had crossed the Andes ten times. He examined the effects of the altitude of the Arequipa Volcano region for the American Geographical Society and made such a careful study of Peruvian earthquakes that he was to be regarded as a high authority on that subject. For his return from Peru, he took the river route to Belem and reached that place only after a journey alone in canoes and by steamer that covered altogether 3,550 miles.

Benjamin R. Whittaker, the coadjutor of Dr. Isaac Coates was a surgeon and physician of marked ability and was as a *bon vivant*, a jovial companion and an all-round good fellow, possessed of such tales of anecdote that his cures were generally attributed to his irrepressible humor rather than to his medicines.

Still another name on the list was that of Franklin A. Snow, assistant to the resident engineer, who was then quite a young man, understood to be a nephew of Colonel Church, and is now a prominent civil engineer of Boston. He has since had quite an extensive engineering experience all over the United States, at the Panama Canal, and throughout Chile and Peru.

Further down the passenger list, we find the names of Mr. Charles L. Moore, who, for the assistant to the contractors, had been appointed American Counsel at San Antonio; Mr. Maurice Mauris, special correspondent of the New York *World*; and Mr. Grandchamp, interpreter. Comparison of names lists three, Thomas J. Cogan, Alfred W. Newton, and George A. Yohe, on the passenger lists of both the *Metropolis* and the *City of Richmond*.

The commander of the vessel was Captain Kelly. He had gone to sea as a cabin boy on a whaling ship at the age of twelve years and remained many years in that service, rising, step-by-step, through dint of industry and merit. In his entire forty years of experience at sea, he had never been in any serious accident of any kind. Yet, despite harsh exposure and hardship, he was gentlemanly, good-hearted, mild-mannered, of a quiet disposition, and a very entertaining conversationalist.

Early in the morning, Willow Street came alive with bustle and excitement and the *City of Richmond* became, for a brief time, the object of the general interest that had been manifested as with the departure of her predecessors. No one could embark on the vessel without a pass. Among the steerage passengers were more than two hundred Italians, the majority fresh from Naples, but recently gathered together from the slums of New York, Philadelphia, and Baltimore. One or two of them were attired in suits resembling those given to convicts at Sing Sing. The *World* correspondent, himself an Italian, saw them as "exhibiting, in shape, countenance, a striking evidence of the soundness of the work theory" and "as going with the courage needed under the necessity of getting a living at last."

The crowd at the wharf increased every minute and almost became riotous. While the officers assisting were busily engaged with preparations for receiving them, singly and in groups of two or three, the excited passengers came on board and, after depositing their trunks, boxes, and bundles, proceeded to inspect accommodations prepared for them. Some local Italians took their wives with them for use as cooks and laundresses at San Antonio. One man, who aroused the suspicions of an officer, was searched and disarmed of a revolver long enough to be mistaken for a rifle and a huge bottle of whiskey.

At 10:30, the steamer moved to Pier 41, South Shore, where the remainder of her steerage passengers came on board. Every available spot in the dock was crowded with spectators. The officers and crew searched all the steerage passengers as they came aboard and relieved them of weapons of many kinds—long knives, daggers, two-edged dirks, and sharpened cooking cutlery of every description.

At 4:00 p.m., there were 423 steerage passengers on board and 27 more expected. Then some of the Italians climbed on deck and insisted on leaving the vessel. Remonstrance was useless, and several were permitted to walk off carrying their trunks and grinning from ear to ear, while the mob on the dock greeted them with shouts of laughter and glee. Fellow countrymen on shore pressed closer to the steamer and endeavored to entice others to leave.

The police then took a hand and drove the crowd back. Thirty minutes later, the vessel cast off lines and steamed down the river to a point below the navy yard, where she stopped long enough to transfer friends of the passengers to the *Rebecca*. An hour later, the police boat *Stokely* put in an appearance and mystified the passengers by closely following the *Richmond* all the way to the head of Green Island Cove where, at 6:00 p.m., the steamer anchored for the night. The Italians were then marched on deck, and there scrutinized by a half dozen city detectives. It soon became known that the object of the search was a notorious criminal, charged with murder, who was supposed to be on board but could not be found.

The formation of messes occupied much of the evening, and early on the morning of the fifteenth, the *City of Richmond* weighed anchor and, without further delay, continued on her voyage down the Delaware and out to sea, whither we shall not follow her. The reader must consider him- or herself as booked for Pará at a later time on a much smaller and less comfortable craft. Blamed for the apparent disregard of his feelings in the choice of transportation and the deficiency in luxury of accommodations, he very naturally claims, as can be promised with confidence, one of the most remarkable ocean voyages that the annals of sea and exploration contain.

XIII. Mackie, Scott, and Company, Limited

If you build castles in the air, your labour will not be lost—that is where they should be. Now put foundations under them.
—Browning

The construction of the Madeira and Mamoré Railway, though the most important and most difficult part of the great project Colonel Church was preparing to undertake, could not be expected to produce the anticipated commercial development until steamboat lines were established on the three thousand miles of navigable water known to exist in the upper Madeira and its tributaries, and until steamers, not subject to the caprice and rapacious charges of the Amazon Navigation Company, were regularly plying between San Antonio and New York City, a distance of about 4,700 miles.

Even before the sailing of the *Mercedita*, an exposition of the deplorable state of our commercial relations with the South American states, made by Colonel Church in the *Philadelphia Times*, led to action by our government and by the government of Brazil, which a little later resulted in the establishment of a new company of steamships, running between New York and Rio de Janeiro and stopping at several intermediate ports, including Pará. The two vessels of this line, *City of Pará* and the *City of Rio de Janeiro*, were known to be of up-to-date construction and were among the best vessels afloat at that time.

Unfortunately, their draft was such that they were forced to anchor in the river two miles below Pará, and the terrible tidal current between the shore and the steamer made the lighterage charges over the two miles greater than the freight charges should have been for the entire distance of 3,200 miles to New York. Sometime in 1877, Colonel Church called upon President Grant in Washington to explain the great need of a survey of the Lower Amazon and Madeira Rivers. The president at once recognized the necessity of the work and promptly issued orders that led to the dispatch of the USS *Enterprise*, under Commander Thomas O. Selfridge, with instructions to map the Madeira and Amazon from San Antonio to the coast.

Next, Colonel Church turned his attention to the great problem of river navigation above the port of San Antonio and endeavored to interest Colonel Thomas A. Scott, then president of the Pennsylvania Railroad, in his plans for the commercial development of Bolivia and Brazil.

Colonel Scott, though much impressed by Colonel Church's description of Bolivia and the great possibilities of the region at the headwaters of the Amazon, was not inclined to grow enthusiastic over the project, the essential part of which was a railroad already mortgaged to Franklin B. Gowen and the Philadelphia and Reading Coal and Iron Company.

Mr. Charles Paul Mackie, who is still living in Englewood, New Jersey, when a mere boy, endowed by love of adventure, had gone to work with Colonel Church and Governor Samuel G. Arnold of Rhode Island. Mr. Mackie subsequently became one of Colonel Scott's secretaries on the Pennsylvania Railroad and, later, secretary of the Pennsylvania and Pan-Handle Companies.

As he was trained under a mastermind to grapple with problems of gigantic magnitude and was intimate with both Colonel Scott and Colonel Church, it was only natural that Mr. Mackie, then young, energetic, and enthusiastic, should be drawn into the discussion of questions with which his residence in South America had made him familiar and finally should be induced to actively participate in the execution of Colonel Church's plans. The circumstances that brought about his connection with the Madeira and Mamoré enterprise, the magnitude of the task assigned him, and the energetic measures taken to ensure its successful accomplishment are here stated in the following quotation from a letter prepared by Mr. Mackie himself:

> Thirty years ago the interior of South America was as remote to North Americans as was the interior of Africa. Colonel George Earl Church, in 1876, was working in Philadelphia, pursuing his wise plans for opening up the whole interior of South America, but particularly the Republic of Bolivia, in the interests of North American commerce and industries.
>
> He had arranged for the building of the Madeira and Mamoré Railway with Mr. Franklin B. Gowen, of the Philadelphia and Reading Railroad, and the

Pennsylvania's Amazon Princess Railroad

Messrs. Collins, then among the most reputable of American railway contractors. Colonel Church invited them to organize an auxiliary company to develop the commercial resources of the extensive region naturally adjacent to his projected railway, and, as I had been previously associated with him in negotiations with the Bolivian Government, offered to assign to me exceedingly valuable rights, under charters and privileges from the Republic of Bolivia and the Empire of Brazil, which he then controlled.

In a way this whole enterprise, with all its vast reaching consequences, drifted to Philadelphia and the natural outcome of the Centennial Exposition of Technology. I greatly doubt whether at any time since that year the full importance to the United States of that state fair has ever been adequately appreciated. With none held since, on this continent or on any other that has contributed a tithe to the development of the United States of that which followed the Centennial.

At the invitation of Colonel Church I had accompanied him to the interior of South America while still a lad in my teens. My father was made vice-president of the National Bolivian Navigation Company, of which Colonel Church was the president, and in which all of the grandiose, but by no means contemporary plans of the latter gentleman centered. By reason of the Centennial Exposition, Philadelphia loomed so large in the eyes of the industrial, commercial, and financial world that Colonel Church chose it as his base of future operations, I was secretary for both the Pennsylvania and the Pittsburgh, and St. Louis railroad companies and acting as aide to Captain John P. Green, then assistant to Colonel Thomas A. Scott.

These were two great, wise and far-sighted men, absolute slaves to work and to the success of that enterprise. Incidentally I may say that the man who will write the railway history of that period, giving initial censure where it is due and praise where it belongs, to big and little men alike, will make a great contribution and do more for the financial solidarity of our public corporations in the United States than any of the muck-rakers who have ever drawn the breath of life, for there were giants in those days.

Colonel Church offered me, with the acquiescence of the Collins brothers, a highly remunerative contract for transporting all men and material required for the construction of the Madeira and Mamoré Railway, to Pará, at the mouth of the Amazon, to San Antonio then the head of steam navigation on the Madeira. In addition, he offered me the monopoly of navigation, commerce, and trade over all the rivers of Bolivia and the immense region drained by them, until the railway could be completed and the National Bolivian Navigation Company was ready to enter upon the full strength of all its extensive concessions.

This proposal meant the establishment of American commerce and transportation on the Amazon River, under conditions which should have prevented its disuse for all time, and the opening up to American commerce of the interior of Venezuela, Colombia, Ecuador, and

Bolivia, the Argentine Republic, and the then Republic of Brazil, with an advantage to our merchants of at least fifty *per cent* in introduction over the British and Germans, then monopolizing a trade of many millions per annum with the territory named. This offer of Colonel Church was based upon concessions from the Governments of Bolivia and Brazil, and was backed by a fund amounting to several millions of dollars, which, subject to the terms of his concession, was then lying at his disposal in the Bank of England. Needless to say, I follow authentic documentation.

To sever my railway connections and dissolve my official relations with Colonel Scott, Captain Green and other officers, who had been uniformly generous in dealing with my shortcomings, required a large wrench, but, when I submitted Colonel Church's proposition to Colonel Scott, he candidly advised me to undertake the work and asked me to associate his son, Mr. James P. Scott, with me in the undertaking. Whence resulted the firm of Mackie, Scott and Company, Limited, and great trouble for every one interested therein.

The story of our purchase of two little river tugs, the *Juno* and *Brazil*, as well as the incidents attached to their voyage to Pará will be related by Mr. Scott for to him, as leader, and to his associates belongs the credit of that most daring and successful adventure. The *Juno* was to have been christened the *Bolivia*, to correspond with the *Brazil*, which had been built as the *Hercules*. Some question or other, as to registration or to special legislation by law, prevented the rebaptism of the *Juno* before she sailed. Indeed, they had great trouble in getting away because at that time the Russian Government for which Mr. Wharton Barker was acting as special agent and representative, was building a number of ocean steamers at Cramp's shipyards, and the local authorities were keenly alive to question about neutrality.

All this was arranged and the tugs succeeded in getting away without serious difficulty, but I noticed that there was an Austrian man-of-war lying in the Delaware at the time, and that her yards were manned, as we headed down stream with the *Juno* at the fore; also that the United States revenue authorities thought it their duty to stop us. The tugs, finally got to Pará, thanks to Mr. Hepburn and his crew and in due time the rest of the expeditionary force followed on mail steamers.

Besides Mr. James P. Scott, who stood back of the whole enterprise, despite untoward and often unintelligible difficulties and disasters, Mr. Wharton Barker and his firm, the late Mr. William Thaw, of Pittsburgh; the Messrs. Levick, and other men, of equal prominence in the business and financial world of Philadelphia, lent their financial support and their earnest personal encouragement. Our relations with the Messrs. Collins were most cordial. Our agreements were carefully drawn. Nothing was left undone that could be suggested by those largely interested in the enterprise, or by our counsel, the Hon. George M. Dallas, of Philadelphia.

To-day, almost a generation later, it is easy enough to pick flaws in all these preliminaries and premises. Then it was the united judgment of the many experienced and prominent men interested, that we were bound upon an expedition which could reflect only credit upon ourselves, benefit upon all concerned, and result in the commercial supremacy of our country over the most valuable trade territory then, and still, unexploited in the world.

It is true the programme was a broad one. We proposed putting high-power tow-boats on the Amazon River, to meet the schooners arriving with railroad supplies at Pará and tow them for 1568 miles, against the strong currents of the Amazon and Madeira, to their destination at San Antonio, the initial station of the railway; to furnish barges, in which to receive laborers and supplies from incoming steamers at Pará and tow them up the river in the same way; to use the empty schooners and barges, returning from San Antonio, in collecting rubber, dye woods, and other products of the Amazon and its tributaries, under concessions owned by Colonel Church and his companies, and consign these articles to Philadelphia for sale there by the firm. So much for Brazil.

As to Bolivia, we were to construct three portable steamers, carry them in pieces around the 240 miles of falls and rapids above San Antonio, and refloat them on the Mamoré River, where we expected to have unobstructed access to various other tributary Bolivian streams, containing altogether about 3000 miles of navigable water, and to enjoy an absolute monopoly trade and transportation until the completion of the railway should enable the National Bolivian Navigation Company to take possession and open the region to general traffic.

At that time, the rubber trade was in its infancy and we knew where most of the rubber was to be found. Quinine bark was worth any price per pound, and the principal source of supply was above the falls of the Madeira in Bolivia. In addition, there were roots of sarsaparilla, dye and hard woods, wool, cacao, sugar, cotton, and all kinds of tropical and semi-tropical products to be had from the region to be traversed by these boats, of which products were to be shipped down the river around the falls, to the firm at San Antonio, and thence to Philadelphia. In return, there were the populations of five or six millions of people, requiring all kinds of supplies and mechanical appliances from the United States. Under Colonel Church's concessions and the contracts made with Mackie, Scott and Company, Limited, these operations, both in Brazil and Bolivia, were absolutely assured for such a term of years as would seem, under intelligent direction, to guarantee the permanent commercial supremacy of the United States over more than one-half of the Amazon basin.

In those days, perhaps because both the United States and some of us, engaged in the undertaking, were younger than now, this did not at all take on the semblance of a pipe dream. Not long afterwards

Stanley did the same thing in Africa and received ample *kudos* for it; yet we helped Stanley materially with suggestions born of our experience.

At all events, after the tugs *Brazil* and *Juno* got fairly away, our firm organized an expedition which was to direct the transportation and commercial development of the territory lying between Pará, at the mouth of the Amazon, and La Paz, in the heart of the Bolivian Andes.

We had the cordial support of both the Brazilian and Bolivian governments in our undertaking, and procured a staff of assistants and coadjutors, as to navigation, transportation, trading, development of material resources, and exploration, which it would have been an honor for any man to lead and which had not had its equal from those days to the present within the limits of the Seven Seas. I only hope Mr. Hepburn has made this all clear. There was Hepburn and Lockwood and Keasby and Dr. Jack Pennington and West and Lawford and Morris and all the rest of them; every man of them game, and most of them fit to the limit for the work they had undertaken to do. It was not a light work by any means, for every man-jack of them took his life in his hands in more senses than one. Anyhow, they all reached Pará, met the tugs, and were joined there by me sometime in 1878.

XIV. The Sea Voyage of the *Juno* and *Brazil*

Like the odor of brine from the ocean
Comes the thought of other years.
—Longfellow

With the Bolivian division of Mr. Mackie's field force, we shall not at present concern ourselves. The urgent need for tugboats to tow schooners and lighters laden with railway supplies and provisions up the rivers to San Antonio made their early departure unavoidable.

Prudence would have dictated sending them down the coast of the United States through the West Indies to some nearby South American port from which they could follow the coast to Pará. The wreck of the *Metropolis* had, however, so deranged the plans of the railway contractors that, even with the relief sent by the *Richmond*, it was known that without power to tow schooners waiting at Pará, the entire force at San Antonio would soon be in a desperate condition for want of provisions.

The direct sea route was, therefore, the only one considered for these little boats. Their exact dimensions were: *Juno* eighty-five feet length, nineteen feet beam, and nine feet draft; *Brazil* ninety-five feet length, twenty-one feet beam, and seven feet draft.

The two boats carried thirty men in all. Mr. Robert Hepburn of Avondale, Pennsylvania, as acting general manager, represented Mackie, Scott, and Co. on board. Captain Cottrell commanded the *Brazil* and Captain Denkin the *Juno*. To those not acquainted with Mr. Hepburn, or who only know the quiet, serious individual who passes under the name today, it would be well to state a little incident that has accidentally come to the writer's knowledge, as an illustration of his character. When a boy of thirteen years, he suddenly disappeared from home. After a long and diligent search for the fugitive, his father finally succeeded in collaring the young man at Camp Carlile where he had gone as a drummer boy in response to the call for volunteers to protect Pennsylvania against General Lee's invasion.

Though only twenty-eight years of age at the time of which we write, he had completed a course of surveying at Princeton; spent some years in constructive engineering work; sailed to Central America, California, and Japan; later visited Europe; and traveled along the entire coast of South America. From a log kept by Mr. Hepburn, which he transmitted to Mackie, Scott, and Co., as well as from letters addressed to that firm, we are fortunate in being able to present a reliable account of a sea voyage, that, so far as the writer is aware, has no parallel in the annals of sea navigation. The newspapers of that date tell us with substance that at 1:00 p.m. on May 6, 1878, after being watched and delayed by a government revenue cutter under the ridiculous suspicion that they were Russian privateers of Lilliputian dimensions, the *Juno* and

William Lawrence Adams

Brazil steamed away from Queen Street Wharf, Philadelphia, on their long voyage of 3,200 miles to Pará. A large concourse of citizens and invited guests had assembled to witness their departure, among the handsome officers of the Pennsylvania Railroad, a number of ladies, and the owners, Messrs. Mackie and Scott.

A party of about fifty or sixty friends and relatives of those interested in the remarkable undertaking, went down the river on the two tugs as far as Wilmington. The crews engaged for this perilous service were apparently possessed with either a spirit of perfect recklessness and indifference to consequences or a supreme confidence that nerve and pluck would carry them through. A luncheon was served on board, and officers and crew alike assumed a forced air of jollity to keep up the collective spirits of the party. Both boats had their bunkers full of coal and their decks covered with bags of the black diamonds, piled two deep. The *Brazil*, with only an eighty-five-ton carrying capacity, started with seventy-nine tons of coal and twenty tons of fresh water, in addition to stores, ropes, etc. The *Juno* was overloaded in like manner, and it was remarked, in going down the river, that the guards of both boats touched the water, whereas, with the boats properly loaded, they should have been two feet clear. Before reaching Wilmington, a leak was discovered in the bulkhead separating the storeroom from the freshwater tank in the *Brazil*. This was repaired as thoroughly as circumstances would permit. At Wilmington, a repeat of the parting scenes usually expected on such occasions caused some trifling delay, but this trying period passed. The captains put on a full head of steam and ploughed the quiet waters of the Delaware

Bay foam until they reached the Breakwater at a late hour and came to anchor for the night. Early on the morning of Tuesday, May 7, both boats put to sea, officers and crew alike determined to succeed in the performance of the dangerous task assigned them and to accept with as much equanimity as possible the only probable alternative—a watery grave.

From this point, we shall continue the story giving extracts from Mr. Hepburn's log and letters to Mackie, Scott, and Co. It is good to recollect that the log was kept for transmission to Mr. Hepburn's employers. This will explain why, in the trying times that followed, the daily entries made therein often assumed an epistolary form.

Tuesday, May 7

Weather very fine. Sea calm. Started straight away for St. Thomas. I went down into the fire room and saw everything going satisfactorily. Commanded them to keep a record by bags of coal consumed.

Wednesday, May 8

A terrible sea on apparently running from quarter of the compass at one and the same time. We have shipped seas by the half dozen and rolled and pitched and chunked about until our heads swam.

We felt anxious last night about the *Juno*, but it came through the ordeal safely and seems to ride the waves very much better than the *Brazil*. Every man seasick or squeamish but the captain. The first one quite sick, chief engineer ditto, two of the deck hands likewise and one fireman absolutely stretched out. The fresh water is all gone except two barrels. The bulkhead leaked so badly we could not stop it. I truly fear we will suffer before reaching St. Thomas. Now using salt water in the boilers. By hailing the *Juno* we find that the fresh water she carries is so putrid and so brackish, in consequence of the salt water running through her uncaulked decks, that her crew is as badly off as we are. On consultation with the captain, decided to issue a bottle of beer to each man, as it became necessary, in place of the defunct water. It is all we have for them to allay the suffering consequent upon the tug's inexcusable neglect, in not testing the work on the tank, before sending us out to endure 1400 miles of unquenchable thirst. The women suffer especially. They come up fairly gasping for water and, strange to say, water is the only liquid that will give them relief. If we had Mr. D——e here, we would drag him astern for forty-eight hours in the hope that such enforced penance would contribute toward his salvation, but I fear even that would scarcely expiate his sins. The water, of course, affects the coffee and tea made with it. A blue outlook! Personally, I have relapsed from squeamishness to a regular attack of seasickness. The weather and wind are fair, but an exceedingly ugly beam sea on, and the *Brazil* seems to take advantage of it to ship all the water possible over her decks. She jumps about in more than perfect accord and sympathy with the rolling, white-caps. Lawford and the captain are the only ones on board who have escaped a severe attack of seasickness.

Thursday, May 9

Please give our combined, co centered and unanimous compliments to Messrs. D——h and E——l, ship's chandlers, and tell them that we sincerely hope, in the course of events, that they may suffer tenfold, and if possible, more, for want of fresh water than we do now, and then may be dependent upon a damndeble compound of molasses and the Lord only knows what else, sold from their stores as Lennett's ale. Three bottles thus far, out of twelve, have proved palatable. This is horrible. No water and dependent upon swill like this! Our friends on the river drank up nearly all our beer and here we are in a devil of a fix without drinkables. The two barrels of water, our only dependence, are nearly as bad as so much salt water. They never were washed out and the taste of the tar and other unknown substances in the water is far from pleasant. Their contents are now used solely and exclusively in the culinary department. Withholding appearance of reason, the men attribute the present epidemic of seasickness to the bad water and the smell of the crude oil which was used, instead of varnish, to finish off the boats with. The ship's chandlers are indebted to the *Brazil* as follows: One wig for the 1st engineer. He inadvertently laid his head against the side of his berth and it stuck fast. Extracted great quantity of hair in obtaining freedom that he sees a wig as a necessity. About ten raincoats. Hang them up or place them for a night against the sides of any part of the boat and they stick fast, only to be loosened at the expense of the buttons, sit down for a minute, lean back, and you must break loose. It is an outrage and a shame and you should deduct accordingly! There was a foot of water in the forecastle this morning and our back cabin had enough to soak everything in contact with the floor.

This is getting serious! Our future assistant engineer was nearly washed overboard to-day. Only the life line saved him. All things considered, both boats are doing well and keep within easy signaling distance of each other all the time. We have been quite a little annoyed by being compelled to slow down or run under one bell for the *Juno*, although, at times, she spurts ahead with a vengeance. She seems to fail utterly in keeping up an equitable and reasonable speed. I will have the water tank of the *Brazil* caulked and repaired and have the supply of water for both boats renewed at Charlotte Amalie, St. Thomas. Will, also, put on board some beer and claret. I wish no repetition of this state of affairs. Feel relieved you will heartily endorse any action that will prevent another trial of endurance like this.

I fear the firemen will give out; they are suffering greatly now. I feel very badly about the state of affairs. You know how men are and they may make desert when we reach port. All our ablutions are performed in salt water and no chance or tub for a bath. We are averaging about four and a half tons of coal per day to date.

Friday, May 10

Had a very rough sea on last night again. Cannot understand the *Juno*. Yesterday she was either pulling away or slowing down all the time. To-day she is going steadily with us. During yesterday's run, as to the strange action of the *Juno*, I overheard from different men on the *Brazil* such remarks as, "That Jenkins must be drunk again," and "Oh! he was drunk when he left and hasn't recovered yet." I must keep an eye on him in the ports as much as possible and Davis (manager) must be prepared for stern but positive action. You failed to carry out your intention of giving me a sealed letter of instructions. It is just as well perhaps. I do not think the contingency will occur, in which the possession of such instruction might prove desirable. There are more things relative to the crew on board that I can much better communicate verbally than in writing. I think decisive treatment the best, but we must wait until they get to the Amazon. They are suffering too much now to enforce discipline, or draw the lines too close. I think it would do no harm to commence with the assistant engineer of the *Brazil* and inform him of the startling fact that he is not supreme on board. We do not propose to allow any of the crew to go ashore at either St. Thomas or Pará except when duty requires. We can keep them busy overhauling and repairing.

The boats are good, but I can swear to the susceptibility of the *Brazil* to every fluctuation or stir in the ocean. Even in this comparatively quiet water where an ocean steamer would be as steady as a rock we cannot keep our hatches open, as the spray spreads over us continually. The *Brazil* is very much of a log in ocean travel. Everything about the boat, inhabitants not excepted, is as dirty and sooty as can ever be imagined. We have all recovered from our seasickness, but feel the want of water very much. The seamen are giving out during their watches. Have had southwest winds ever since leaving Cape Henopen. Can't imagine where our northeast trades are as we have been in their latitude for two days.

Saturday, May 11

Both boats steamed alongside of each other to-day until about 6 P.M. yesterday, when the sea became, for the first time, quiet and smooth, we thought it the proper time to coal the *Juno* as per instructions from you before leaving Philadelphia. We imagined that she needed it and subsequent information proved our surmise correct, though the engineer informed the captain to the contrary. We had understood Captain Denkin, when within hailing distance the day before, that he only had his bunkers half full. We whistled to attract attention to us and ran to meet him. When within hail, we asked about his condition as to coal, and, being somewhat miffed, answered that he had sufficient to reach St. Thomas, I told him the *Juno* had better precede us and we would blow on her quarter. Captain Denkin called back, "Do you want me to go for St. Thomas?" I replied, "Yes," and away he went, hauled off to the eastward and ran out of sight. For a time, we could discern his lights and smoke, but after a while we lost him altogether. I think there must be some misunderstanding. Captain Denkin may have construed our offers of coal as an insinuation that he was unable to keep up with us and thought we were "rigging" him. We had some difficulty with our jets, which were plugged up with the salt water we were using, and lost some four hours' time remedying the trouble. I propose to have a very distinct understanding about the running of the boats from St. Thomas to Pará. No more of this foolishness! They must keep up a good regular speed of about nine knots and keep to that like clock work.

It will be most consistent with the principles of safety, time and economy. The *Juno* will precede with the *Brazil* on her quarter, or slightly astern. This being without fresh water is terrible and can only be appreciated by those who have had an experience similar to ours. I do not wish to exaggerate or be pathetic, but, to see these firemen come up from below after every turn at the fires, lie down on the deck and gasp is very trying to ordinary humanity. I fear the influence such a truthful tale may have on a human. When we reach St. Thomas, will not stint in making the boats secure against a second experience of this kind. They must leave that place in the best condition possible.

Sunday, May 12

The lights of the *Juno* were just perceptible on our weather quarter last night about 9.30. Have seen nothing of her since; not the slightest indication of smoke or boat, as far as we can detect on the horizon.

We are pushing for St. Thomas as rapidly as possible, entirely disgusted with all we have on board to drink and our stomachs craving relief. It is impossible to do justice to the misery and suffering of the firemen, upon whom we are dependent for our arrival at St. Thomas and for a supply of fresh water within a reasonable time. They are the very ones most likely to succumb and I do not know what we could do in a crisis, an extremely probably contingency. We could all help at it, but how long could our combined endurance last when men trained to such work give out?

We have been dealing out a mixture of ale and vinegar in lieu of water, and a despicable mixture it is. The ale supplied by the chandlers is about out, two bottles only left. I believe every one from captain to cook is half sick from the use of it. Two or three bottles only were at all palatable or rather endurable, but necessity is a hard master. I wish again to hurl condemnation and abuse on the head of all responsible for willfully and deliberately sending a boat out for a 1400 mile voyage, entirely out of sight of land, with an untested water tank to depend upon almost for existence. I hope sincerely you will hold him to an effective and lasting account and do the same with Messrs. D——h and E——L for their retched molasses compound.[15]

If you send another boat out, for Heaven's sake see that she is perfect in every detail. It may save life and will certainly make it possible to exercise a reasonable control over the men.

There is an ugly sea on, which contributes to the general despondency on board. I will put on water casks, in addition to repairing the present tank at St. Thomas. No more of this thing!

We cannot write at all on board and have not been able to do so since we started. I am simply scribbling notes, which I will copy, if I every reach land. Captain and Lawford at last complain of squeamishness. Don't wonder! It is enough to try a copper-smith's stomach! A very ugly sea, and wind east-southeast.

Our cabin aft regularly receives about six inches of water each day and night, saturating all our wearing apparel. This *is* fun with a vengeance! I wish Davis was in charge of his own department. I feel that he owes me a line-long debt of gratitude. Jove, couldn't he turn himself inside out and growl! This is the first boat I ever traveled on, in which, for seven consecutive days, with the exception of a very few hours, you could not move without clinging tightly and affectionately to something. Came near losing another fireman overboard to-day.

---

[15] Mr. Hepburn doubted the justice and propriety of publishing any names, with probably undeserved reflections upon certain Philadelphia firms, although they were deemed essential to illustrate the high nervous tension mankind was laboring under at the time.

The fireman Joe, instead of making the best of a bad state of affairs, growls and whines persistently and incessantly. Gave out before his turn expired and had to put another man in his place. The other men are doing well, but feel bad and are suffering intensely.

Monday, May 13

We broke down at 4 P.M. yesterday. The key from the cut-off cam fell out. The cam, slipping on the shaft, brought the cut-off valve stem in contact with the link, breaking the guides for the cut-off, bending the valve stem and two eccentric rods and twisting the reverse rod forward. Just here I wish to add another charge to Mr. D——e's account. We have discovered, in repairing, that the key which caused the trouble never fitted in its place, having originally, at the shop, been made too small and lined up with a piece of tin (Lawford). It was an outrageous piece of work.

I called Mr. Bellvou, chief engineer, to the Captain's room and received the following: "The key to the cut-off cam came out, broke the guides for the cut off and bent the valve stem and eccentric rods. The cause of the trouble was that the key was too loose and not fitted right. I put in four pieces of scrap tin and it wouldn't hold in its place. The fault was in building the engine." This is a verbatim report of the explanation offered and to it was added the factual statement that "This cam drives the feed pump, sump pump, circulating pump and independent cut-off." We heard a terrific crashing, hammering and jumping in the engine room and rushed down to find the above described condition of the machinery. There were some seriously expressive faces looking at each other as each one seemed to mentally accept the terrible situation that threatened to confront us in case the break could not be repaired. Out of drinking water, two hundred miles from the nearest land, out of the track of vessels, out sails about as much use as a fifth wheel to a wagon and no other motive power, truly a desperate state of affairs! One member of the firm suggested to me before leaving that if anything went wrong with the engine we could *sail* into port. Good heavens! Unless in a gale of wind on her quarter, she could not make twenty miles a day. We had a strong wind blowing when the accident occurred. We headed her to make it a beam wind and she made by the log just about half a mile in two hours. The sails are useless except to steady her, or to slightly increase the low speed of a crippled engine, when the sea is too rough to risk a voyage in our little row boats.

Captain Cottrell and I went up to the pilot house to talk the matter over. I told him, if worst came to worst, I would take any officer or man who could steer and manage a little boat in a heavy sea, take one of our boats and endeavor to reach Porto Rico or St. Thomas and from one of these places send a vessel to rescue them. That indicates how desperate we were. The exhaustion of our water supply was the principal cause of anxiety and fear.

We went below again to find them trying to straighten out the bent parts of the engine on an anvil, after subjecting them to the heat of the furnace. We watched each blow of the sledge as it fell on the rod with most intense interest, feeling that life and death depended upon every stroke and knowing that, if it broke, we were almost hopeless and helpless.

Owing to the ability of the engineer, we managed to get the damaged pieces into sufficiently good condition to go on. We were stopped three times by the key coming out, but, as they were closely watching, no further damage resulted. That night there was a continuous series of terrific squalls, accompanied by rain. We rolled, pitched and jumped about so that we could not stay in our berths, nearly lost our sail overboard, had to take it in to prevent its being torn off.

If you could have seen the poor firemen and others, rushing about with cups and buckets to catch every dribble of water, as it ran off the roof into small boats from the torrents the heavens were pouring down upon us, you would have felt your heart thump as we did. The first officer spread a sail and caught all he could. We drank most heartily and thankfully. It was a temporary relief. We arrived at St. Thomas and came to anchor in the harbor of Charlotte Amalie at 5.30 P.M. The *Juno* had beaten us by about 50 minutes, making the run in 6 days, 19 hours and 30 minutes from Philadelphia.[16] I at once ordered a supply boat alongside and the men gleefully reveled in good ale.

---

[16] This allows for a loss of time, while anchored at the harbor breakwater all night, of seven hours and forty-five minutes.

From a letter written at St. Thomas by Mr. Hepburn to Mackie, Scott, and Co., the following excerpts are made.

> The above log is taken from hastily scribbled notes in lead pencil. I have cut down and shortened the account of the voyage, and in many instances moderated the language and tone of the original, as I feared you might think hyperbole a weakness with me. The notes were written under the pressure of intense feeling caused by the perils to which we were subjected and, if they even yet breathe a virulent or defamatory spirit toward certain persons unjustly, I can only offer our discomfort and danger as an excuse.

That man Howard, fireman, came very near dying during "the water trial." He had been very seasick and thirst so aggravated his sufferings that I thought, as we rounded St. Thomas, that it was a case of canvas bag and lead.

The *Juno* lost a great deal of coal in the Gulf Stream. I believe the men threw some of it overboard. I do not wonder much, as it was a trying night. She arrived in this port with about one ton on board. Her seams leaked badly. Coal and everything else soaked. Her crew, like ours, nearly all seasick. Her engines were in poor condition and pounding badly.

In short, both boats and occupants arrived badly in need of rest and repairs.

I wish to state here, that, of all the places I have visited or heard of, this is the worst for a stranger to attempt to transact business satisfactorily and expeditiously. To hear business competitors rate each other, one would think his only choice was between a notorious villain and a confirmed and convicted scoundrel.

You can scarcely imagine the pertinacious harassment to one arriving unconsigned, unadvised and totally ignorant of place and people. The "to-morrow" affliction is as serious an epidemic in this place as in any I have ever visited. To get a trifling piece of work done, one must either personally or by proxy, hang to the person who promised to do it, with bulldog tenacity and even then, be driven nearly frantic with impatience at the slow and deliberate movements of the workmen. Our repairs in the United States, even at D——e's, could have been done in forty-eight hours, but here they affirmed and swore that it was an impossibility to complete them before the 21st. I raised such a row that, by repeating your Philadelphia method of giving personal attention to the work, we expect to leave to-morrow, Friday. I hope the excessive profanity I have indulged in will be forgiven me in consideration of the cause. It is no amusement, after the fatigues and dangers of our ocean voyage thus far, when all are sadly in need of rest and recuperation, to be compelled to get up at 5 A.M., start out after coffee and keep going in this warm climate until 9 P.M., doing one's writing between times and late at night, rushing this man here and that one there, sitting or standing alongside of an anvil in a blacksmith shop, or a lathe in a machine shop, demanding, coaxing and cajoling in an endeavor to expedite the work. You can hardly expect to hurry a man when there are only two in the place capable of doing your work, and one of them so rushed with other business that his service cannot be secured on any terms ... Our chief engineer, Mr. Bellvou, sent enough work to the shop to keep us here for ten days, but I vetoed most of it.

Many passages in Mr. Hepburn's letters indicate that there was serious difficulty in maintaining discipline among the crews of the two boats and a reluctance on his part to enforce rules and regulations too strictly until the crews were out of reach of the American consuls at St. Thomas, Barbados, and Pará, who might attach undue weight to their tale of past suffering.

He writes:

> They are a bad lot—some of them—and impudent in proportion ... Seem to consider this a grand independent communistic lark ... If it had not been for

> the water business, we would put the screws on them now ... Had to put a man in chancery (irons) last night on board the *Brazil* for profanity and impudence ... Heard from police headquarters that two of our firemen, while on shore, stated they would "make it warm" for us at sea. Think it was drunken nonsense, but will see that irons are ready for them ... I have, in several instances, been forced to show my bristles ... I am trying to inaugurate some discipline here, but we cannot do much until we reach the Amazon. The men had a strong case against us on this water business and we have shown some leniency therefore.
>
> Many annoying things have occurred, that it would be puerile to write about, but which add cares to the responsibility that, under the most favorable circumstances, would be weight enough for any one to carry.

On the evening of May 16, Mr. Hepburn, as an acknowledgment of exceptional courtesies extended by the captain of the port and others, gave a farewell lunch on board the *Brazil*, the guests being the consuls of the United States and Brazil, the captain of the port, and a few others.

In another place, Mr. Hepburn says:

> As you perhaps know, this island contains two structures, known as Bluebeard's and Blackbeard's Castles, which tradition tells us are relics of those days of piracy and rapine when buccaneers sailed the Spanish Main. Though very busy, we took time, one beautifully clear moonlight night, to visit one of these curiosities

and were hospitably entertained by the warden, who is also the captain of the port. The two castles are round stone towers, situated about two thousand yards apart, high up the rugged side of the mountain which rises somewhat abruptly behind Charlotte Amalie. They are known to have been erected by the Government in 1688 as a defense against the pirates, who are said to have subsequently seized them and made them famous. Certain it is that they have long borne the names of their supposed occupants, Bluebeard and Blackbeard, two brothers, who are believed to have used them as places of security, from which they could command an extensive view of the ocean and watch for treasure-laden vessels. Blackbeard's Castle is the taller of the two and has four encircling rows of portholes, from which small cannon could be fired, while Bluebeard's Castle has only two. Both were out of reach of the ancient artillery used on shipboard at the time they were erected, and well-manned, were capable of successfully resisting any small land force likely to be sent against them. On the ground floor of each tower is a roomy dwelling in which the pirates are said to have lived when not employed in the pursuit of vessels. There, they are said, to have kept their female captives and with their own hands to have murdered the older ones for mere amusement, or to make room for others younger and more attractive.

Bluebeard's Castle has well endured the storms of nearly two centuries, and as we looked out upon the mountains clad in semi-tropical verdure, surrounding, in the shape of a horse-shoe, the landlocked harbor, dotted with ships form every part of the glove, upon the ancient town, with its quaint structures below us and upon the broad expanse of sea, all bathed in the mellow light of the moon, our minds conjured up such romantic visions of the past that, like the natives, we were inclined to resent the exhibition of any iconoclastic spirit toward the tradition of the place.

Returning once more to Mr. Hepburn's log, we find under the date of Sunday, May 19, 1878:

We left St. Thomas about 2 P.M. yesterday, sighted and ran around the Pussey, Jones & Co. side-wheel steamer *El Progresso* beyond the entrance of the harbor. She left Wilmington, I believe, May 4th. We were in St. Thomas five days less a couple of hours. Captain Cottrell and Bellvou, chief engineer, had some words and exhibited some temper about a question of authority on board the *Brazil*. I had a talk, first with the captain and then with the chief. I told them both in effect that the captain must be the authority on board, there must be a head, or there would be chaos in our organization in a very short time. In all detail about the engine room and boilers, Mr. B. must shoulder the responsibility. The captain's supervision must be general. Any appeals must be made through the captain to the one above him, no tittle-tattle and no "running" would be permitted. It must be a matter of gradation, each one to shoulder his own share of responsibility and stand or fall by his individual record. I have tried to inaugurate some system of discipline, a respect from each to his superior and to altogether eradicate the tugboat free and easy system. It is a difficult task, but I think the fight is over. The tugboat idea, that each and every man was as good and equal in authority to the captain or any one else, we have endeavored to overthrow and suppress. In a degree, I believe we have succeeded. Firmness and persistency will fully accomplish the final obliteration.

During Mr. Lawford's watch, I am sorry to report for he has been exceptionally faithful and is always of great confidence, though careless at times, we had a slight difficulty caused by the fireman on watch neglecting to oil a guide. The bearing soon became red hot and away went gib. No damage done except 2 hours' delay. Captain Denkin, missing us, came back to ascertain the cause and whether we needed assistance. Captain Denkin is, in my estimation, one of the very best men you could have secured for this venture. He is determined, able, economical and honorable.

Mr. Lawford's difficulty was misplaced confidence. It was, perhaps, a cheap experience. I sent for the fireman who caused the trouble and suggested to him that any more occurrences of the kind would not only subject him to loss of standing with the company, but would also entail a series of other unpleasant consequences, the least of which would be irons and a diet of bread and water. I thought that perhaps this was one of the methods of revenge reported by the St. Thomas police, and made up my mind, in the next instance, to make an example of these petty would-be muitineers. Everything now seems to have settled down to a system, but we keep a sharp lookout for trouble.

Monday, May 20

We have had pleasant weather overheard, but a very rough, rolling, tumbling sea underneath, and high winds. We shipped a great deal of water all day yesterday and last night the *Brazil* acted as if she was suddenly attacked with Saint Vitus's dance. All well on board, but very uncomfortable. If Mr. Davis does not always have a boat at my service whenever I go within his domain, I will consider it very shabby treatment in return for the service I have done him by putting myself in his place for this trip. The men think the company should pay them for their loss in clothes, caused by the adhesiveness of this stuff on the side of the boat, and by the forecastle being semi-furnished and normally half full of water. As I might be a claimant were damages awarded, I will not plead the justice of our cause. We would like to get hold of D——e and drag him astern for twenty-four hours.

Wednesday, May 22

Arrived in Barbados about midnight on the 20th. Took on 12 tons, 500 lbs. of coal on the *Brazil* at $9.25 per ton. Water was a necessity again. We had very bad weather Sunday morning last. Found on arrival that one of the Royal Mail steamers to St. Vincent had extended her schedule time of 10 hours into 24 hours on account of stress of weather. They were much surprised at our time of 2 days and 10 hours from St. Thomas and would scarcely credit the fact of our having come from Philadelphia. We have created quite an excitement thus far. The undertaking it denominated *"another d--n Yankee experiment to show their natural recklessness."* Our 2 days and 10 hours' run included the two hours' delay previously noted.

While at Barbados, we were boarded by an officer of a large English warship lying in the harbor, who was curious to know what such diminutive craft was doing so far away from home. He was a young, fine looking fellow and seemed very much interested in inspecting the boat. After he had finished and glanced over our log book, he turned to me with a quizzical look and said, *"Lives must be plenty in the United States to send men to sea in vessels of this size. You seem determined to keep up the national name for reckless adventure."* We treated the matter coolly and indifferently before him, but would willingly have transferred the privilege of keeping up the national name to some one else for the remainder of the trip. It was just about one year before that Captain Symmes[17] left Barbados for Pará with some such craft as we and was never heard from afterwards.

This story was retailed to us, as you may imagine, with all shades of high coloring. While it did not create nervous debility, or frighten us very badly, it

---

[17] This was the father of the Captain Symmes who was a passenger on the *Mercedita*. The writer's impression is that the father attempted to sail in a floating dock to some South American port.

was far from pleasant, cheerful or encouraging under the circumstances. I had promised a half holiday on shore to the crews of both boats after the coaling was finished, but upon hearing the above story, feared the effect it might have and decided to sail at the earliest possible minute.

While breakfasting at the hotel, on the morning of the 21st, the proprietor came into the room followed by what at first appeared to be a startling apparition, a tall, gaunt, death-hued semblance of a man. Following so closely upon the tale of disaster it was uncanny. On being introduced he took me to a window and pointed to a large full-rigged ship lying in the harbor and stated that she was his vessel, that he had come from Santos, Brazil, that yellow fever had broken out on board, that several of the crew had been buried at sea, that the others were all weakened, that he himself was only convalescing, that there was a fair wind outside, but a head wind in the bay and he had been unable to get out for some days past.

I anticipated him by saying "You wish to be towed out." He promptly answered "Yes," and asked what it would cost. I said, "You are an American and so am I. You are in distress. Have your hawser ready at three o'clock this afternoon and you will be taken out and put upon your course. There will be no charge." Tears filled the poor fellow's eyes. He wrung my hand and went out as if the elixir of life had been infused in his blood. An English ensign, who was present, remarked, "Do you intend to try to tow a loaded vessel of that tonnage out of this port against wind and tide with one of those little boats?" I replied, "We're not going to *try*, but are going to do it." "Well," he said, "if you succeed, we will give you a send-off from our ship."

Those familiar with the cumbersome English tug boat of that day will readily understand the skepticism of this young officer as he scanned the low-lying American craft. I had intended taking the ship out with the *Brazil*, the larger of the tugs, but in view of the doubts of success on all sides decided to use the *Juno*.

At the appointed time, the wharf was crowded with excited people of all classes and even on the ship in the harbor, an unusual stir was apparent. The *Juno* steamed over, made fast to the hawser and, everything being ready, started down the bay at a speed that astonished the on-lookers. The *Brazil* weighed anchor and followed amidst roars of applause from harbor and shore. As we passed the English man-of-war, her decks and rigging were crowded with officers and men and we were greeted with hearty cheers, shouts for good luck and success, and a salute with their guns. After rounding the point, the *Juno* dropped the ship hawser and, at 6 P.M., accompanied by the *Brazil*, proceeded on her voyage to Pará. There was not a man, woman or child, afloat or ashore, outside of the American vessel and our boats, that believed we could move the ship one hundred yards for her anchorage.

Thursday, May 23

The *Juno* dropped quite out of sight behind this morning. We ran back to find out the cause and see if help were needed. Found our engineers had borrowed a wooden gib from the *Juno* in Barbados and had failed to return it. We remedied the trouble and went ahead. Are driving against a head wind and sea. We have had to contend against this combination almost continuously since leaving Cape Henopen, causing great discomfort, a considerable reduction of speed and a large increase in fuel consumption. We have yet to experience one comfortable hour on this boat. Our state-rooms are being continually drenched with water and our clothes floated about promiscuously. Our table was upset and knocked clear out of the sockets for the legs this morning. It made a mess of everything. We are all more or less bruised and battered from being thrown about. Captain Denkin, in telling us, complained of inability to sit down any more until healed. Thank heaven, we are nearing the end of this ocean voyage. There are very few on the two boats that could be induced to try a similar experience again. Coast down? Yes! Straight-away? No! The wind still continues south-easterly. Where are the northeast trades?

Friday, May 24

Nothing of note to-day except our pitching about against a head wind and sea. We had a beautifully calm sea this morning, but it did not last. There must be a strong current against us, as our log registers 225 or

235 miles and our sights only give 188 miles. My notes are made in a very desultory manner. Can neither read nor write, but can growl most successfully. Had several very bad squalls last night. Blew and rained very hard. The current seems to be about one and a half miles per hour against us. Sea is becoming calm again.

Saturday, May 25

Had three very bad squalls last night with a heavy downpour of rain. I always fear the roof of the pilot house, or house and all will go in one of these fierce squalls. The superstructure is not built for such ordeals. The sea is now as calm as a lake and we appreciate the change, I assure you. Our faith is, however, weak as to continuation, but we still hope. Even the swell of the ocean rocks the boat like a leaf. Hailed the *Juno* yesterday about coal. They think they have enough to reach Pará, though I have little confidence in Mr. Jenkins as chief engineer. Either we were deceived in the capacity of her bunkers, or she averages six to seven tons of coal in twenty-four hours. We are keeping all in good spirits with anticipations of an early termination of the salt water portion of our voyage. I would never try another experiment like this, unless allowed to coast down and run into ports at will, allowing six weeks or two months to accomplish the voyage. I think we all realize more fully the risks, chances and dangers of the experiment in retrospection. Should other boats be sent out, I think the chances even whether they would ever reach Pará, if they came by the same route. There are many hundred dangers to be encountered, which are only realized when one is confronted with them and the success of such an undertaking is largely a matter of chance. Every one with whom I have conversed seemed to consider us crack-brained dare-devils, and I must confess there was ground for such opinion. You may think we are inclined to exaggerate the dangers of our trip, but everywhere we have met with rather incredulous smiles at the very idea of our having come from Philadelphia, or looks of wonder and amazement. Possibly, our own feelings are in some degree influenced by seeing ourselves "as others see us." We are having an average of eight to twelve fierce, rainy squalls every twenty-four hours.

Sunday, May 26

Squalls still continue. Offered *Juno* our hawser if she were nearly out of coal. It was refused. Captain Denkin says they have enough coal to last until to-morrow. Current set us 45 miles to the northeast during twenty-four hours. This is annoying and irritating, as fuel and time are important considerations right now.

Monday, May 27

At 10 A.M. the lead brought up river mud in nine fathoms of water. Are evidently off the mouth of the Amazon and about seventy miles from the lightship. There has been grave misrepresentation or ignorance upon the part of the previous owners of these boats in regard to the amount of coal consumed by them. The *Juno* continues to average over six tons per twenty-four hours, and the *Brazil* nearly six tons. We took the *Juno* in tow last evening, May 26th. She whistled us to come to her and we found she had only half a ton of coal on board, though supplied with thirty tons at St. Thomas, and twelve and a half tons at Barbados. The *Brazil* coaled with thirty-eight tons at St. Thomas, and seven and a half at Barbados. It is getting to be close work again. A gale of any duration would leave both boats out here at the mercy of wind and wave. If you send out another boat, send her down in sections on a steamer, or coast her down. Taking all these chances with small coal bunkers, the odds are against you. Our hawser parted about 2 A.M. We got up our new line and attached it. We move along smoothly now. There seems to be little difference, so far as we can perceive, in our speed with the *Juno* in tow, though, to be sure, they disconnected the machinery from the shaft, so that it might turn easily. We were up all night keeping a sharp lookout, as the current is very swift here. Besides, being handicapped with the *Juno* increases our anxiety. Coal low and altogether an uncomfortable prospect. Captain Cottrell is certainly doing his duty and keeping a keen eye on every man to make time and prevent accident. It is very apparent that he is worried and anxious. The squalls continue, blowing furiously and raining hard, driving us under cover. Every one on board is keeping a sharp lookout for a glimmer of light from the lightship at the mouth of the river, though we are yet some hours distant from it.

Captain Denkin called, when within hearing distance last night, that we should leave him at the light-ship, go up to Pará for coal and then return for him. Time is too important a consideration for such actions as this. If we have coal enough, both boats must go up. The chief engineer just reported only seven or eight tons of coal on board. We are in a very serious plight if such is the case. A forty-five mile current against us, nearly two hundred miles from Pará, *Juno* in tow and only seven or eight tons of coal with which to reach our destination and complete an ocean voyage of 3200 miles. Our fresh water tank became brackish again after leaving Barbados and we have only two casks of fresh water on board. Here we are again facing a possibility of drifting about until some vessel picks us up. I have given the engineers in every instance a margin of three to four days on the running time to estimate their coal consumption by, and here we are, one boat out of coal, the other reported with an insufficient supply to carry her into port, and our sails absolutely useless. A man must be omniscient and ubiquitous to take charge of an expedition like this! Bah! I am disgusted! We will stick it out and stand by the boats to the last. I ordered the chief engineer to bag all the coal he has, give so many bags to each fireman during his watch and have them make every effort to use it as economically as possible. I close to-day with a gloomy prospect ahead. If we are brought into port, the salvage will eat up the first year's profit. I feel disposed to raise a row in some quarter, but don't know where to begin. The sun will set to-day upon a blue lot of men.

Tuesday, May 28

Reached the light-ship at the mouth of the Pará River at 7.30 A.M., 21 days, 17 hours and 30 minutes from Philadelphia; actual running time, 15 days, 19 hours and 45 minutes.

The engineer reported, after he had filled some 40 odd bags with coal, that there were probably 13 or 14 tons of coal on board. There was great rejoicing when this news came up last evening. It seems that the bunkers extended farther forward than he had supposed. We brought Captain Denkin on board last night for consultation as to the best course to pursue when the coal was exhausted, but this discovery makes our plans unnecessary. The experiment is now an assured success. Gave the *Juno* three tons of coal at 8 A.M. Pilot now on board. We are off for Pará! Anchored in the river in front of Pará at 6.30 P.M., *Juno* about 7 P.M.

Here, as elsewhere, we are congratulated by every one, by people on shore and by officers of the United States steamer *Enterprise*, upon our success in accomplishing what has never been done before and is not likely to be attempted soon again. It makes me feel grave to look back, but I have no desire to seek applause, or appear in the role of an applicant for the laurel crown.

In order that the reader may better appreciate the feat performed by these little river tugs, the writer has carefully compared their actual running time with that of ocean steamships, which traversed the same general route during the year 1878. No allowance has been made for delays at sea and no deduction has been made except for time spent at anchor in port. It must be remembered that the *City of Rio de Janeiro*, besides being a perfectly new steamer, had a high wind and a strong current in her favor, which operated to retard the other boats.

Philadelphia to Pará

Mercedita	25 days,	13 hrs.,	30 min.
City of Richmond	21"	17"	50"
Juno and *Brazil*	16"	6"	15"

Pará to Sandy Hook

City of Rio de Janeiro	10 days, 4 hrs., 55 min.
USS Enterprise	21 days approximately

After reaching Pará, Mr. Hepburn, with his usual energy and efficiency, at once turned his attention to the work of towing vessels laden with railroad materials and provisions to San Antonio. No one can properly appreciate at the present day, the vexation, difficulties, and dangers involved in this work. Though his letters are replete with information regarding the 1,568 miles of river navigation, the character of the people he encountered, and their methods of transacting business and with descriptions of the country through which he passed, it is not the writer's intention to narrate the incidents attending a second trip up the Amazon and Madeira. We will, however, so far anticipate succeeding events as to say that thirty days after Mr. Hepburn landed at Pará, when sickness was everywhere prevalent among the men at San Antonio and on the river above, when there were no medicines, not even quinine, for the sick; when provisions were practically exhausted; when deaths were of almost daily occurrence, the tug *Juno* suddenly appeared at San Antonio with Mr. Hepburn on board, the schooner *D. M. Anthony* in tow, and the first adequate supply of provisions and medicines that had as yet reached the sorely tried inhabitants.

XV. Events at San Antonio

How strange are the freaks of memory!
The lessons of life we forget,
While a trifle, a trick of color
In the wonderful web is set.
Set by some mordant fancy,
And spite of the wear and tear
Of time or distance or trouble,
Insists on its right to be there.

—James Russell Lowell

Although the *Mercedita* reached San Antonio on March 7, the absence of proper landing facilities made the unloading of heavy machinery and railroad iron so tedious that she had not finished discharging her cargo on the eighteenth of the same month, when our narrative closed at the end of the tenth chapter.

By March 20, Corps No. 4, under Captain Stiles, had completed the preliminary survey for a distance of two miles below San Antonio. Corps No. 1, under Mr. d'Invilliers, in conjunction with Corps No. 2, under Mr. Byers, had finished the preliminary line from San Antonio to Macacos, a distance of about five miles and from the last mentioned point, and Corps No. 3 was pushing southward. Topographical plans had been prepared in the office, showing the projected location of the railway to Macacos. The headquarters corps were marking this location on the ground and staking out work for the greatly increased force of laborers constantly expected. The old sawmill, left by the English, had been placed in running order and was supplying lumber of all sizes for building purposes.

Such progress at home would have been considered far from satisfactory, but under the conditions prevailing at San Antonio, it was regarded as a subject for congratulations. Every inch of the line had to be cut through almost impenetrable vegetation. Quite frequently, immense trees, after being completely cut through at the base, would continue to stand erect, held firmly in place by the network of vines that tied their tangled tops to everything standing within a radius of fifty feet. No idea could be formed in advance of the nature of the ground to which the preliminary line was leading. The engineer's first knowledge of a high hill or a deep valley in front of him was obtained when his axmen disappeared among the treetops or sank out of sight below him. Surveying through these forests was like working by the light of a lantern at midnight. The rainy season had not yet closed, and the field parties were regularly drenched every afternoon. Stinging and biting ants, in countless numbers, were on every bush and tree and made life on the line one continuous torture.

The provisions sent from Philadelphia, even when abundant, were unsuited to the climate and were put in such bulk as to render transportation on the river and to camps in the interior extremely difficult. Even such inadequate supplies were by this time practically exhausted. A moldy sea biscuit and a canteen of water constituted the usual lunch on the line. Occasionally, a monkey or a parrot was killed by someone, but such luxurious fare was very rare.

Captain Stiles's party had been detained at San Antonio, because of lack of camp equipage. When the arrival of the *Mercedita* supplied this deficiency, there was no possibility of furnishing him with a sufficient supply of provisions to justify moving his corps to San Carlos, the initial point of his contemplated operations on the river above.

In consequence of this state of affairs, it was decided that, as soon as the *Mercedita* was ready to depart, James T. Brown, chief clerk, should return on her to Pará and there make the best arrangements possible for the purchase and shipment of provisions to San Antonio, it being understood that, if on the way he met an American vessel coming to our relief, it would be unnecessary for him to carry out his mission. What we were to live on during the four weeks' time that must elapse before Mr. Brown could be expected to return, was an unsolved problem. We had made existence to some extent endurable by private purchases at the little native store, but the proprietor was not at that time prepared at any price to entirely subsist such a large number. It was, therefore, with gloomy anticipation for the future that our half-fed men at headquarters sought repose in their hammocks on the evening of March 22 and far into the night kept up a discussion of the situation, accompanied by expressions of opinion as to the best possible course to pursue, should it become impossible longer to remain at San Antonio.

About six o'clock on the following morning, some of us

were startled by the sound of a whistle in the distance. Going to the little railing that surrounded the second floor of our headquarters building, where a view of the river for two miles below could be obtained, some of our men exclaimed, "It is an ocean steamship!"

Field glasses were then brought into requisition, and a further exclamation, "She carries the American flag!" emptied every hammock more expeditiously than the most violent earthquake could have done. Never did the Stars and Stripes create such boundless enthusiasm. From the headquarters building, rifles and revolvers were fired to notify those in camp of our discovery. Men, delirious with joyful anticipation of news from loved ones at home and the prospect of continuing their labors under more auspicious circumstances, were hugging each other like schoolgirls, singing, dancing, and yelling like incurables in some asylum for the insane.

The breakfast of cornbread and sugarless coffee was entirely forgotten. From headquarters and the various camps, the men all gathered at the steamboat landing, where at 6:45 a.m., amid the cheers of the inhabitants, the steamer *City of Richmond* came to a stop and ran her lines ashore, thirty-six days, sixteen hours, and some few minutes after leaving Philadelphia. It only required a few minutes for us to discern the tall form of Mr. Thomas Collins on board, and even before the passengers could come ashore, questions were being asked and answered in regard to the fate of those on board the *Metropolis*. Then for the first time, we heard the entire story of that disaster from survivors who had gone down on the *Richmond* and from letters and newspapers in the mail she carried.

From the passengers, we learned that the voyage had been

marked by few incidents worthy of mention. On leaving the Delaware, the captain had, as he supposed, steered straight for St. Thomas. At 4:00 p.m., on February 21, a dim coastline could be seen in the distance, but soon after, a heavy fog settled on the ocean, and the captain deemed it advisable to blow off steam and come to a stop for the night, supposing the vessel to be only fifteen or twenty miles distant from the Charlotte Amalie.

The heavy fog, accompanied by a drizzling rain, continued until 9:00 a.m. on the twenty-second, when clearing weather revealed the fact that the vessel was far out of her course and only a few miles distant from the northern shore of Porto Rico, then distinctly visible. This course was followed all day in a westerly direction until about sunset, the passengers enjoyed a distant view of the city of San Juan. From there, the captain once more laid his course for Charlotte Amalie, where the *Richmond* came to anchor at 9:00 a.m. on the twenty-third and remained until noon on the following day.

The mouth of the Pará River was reached at 1:00 p.m. on March 6. There, the vessel lost thirty-nine and a half hours at anchor or cruising about in search of a pilot. Two boat crews were dispatched for information to a distant lighthouse on shore and, for some unknown reason, failed to return but subsequently reached Pará after great suffering and a pull of more than one hundred miles, much of the distance against a strong current. A coasting steamer, bound for Pará, finally supplied the *Richmond* with a pilot. She reached that city at 10:00 a.m. on the ninth, and there remained until near midnight on the eleventh.

On the morning of the nineteenth, one of the passengers, John Scanlan, a young man, twenty-three years of age, died of pneumonia. This was the first death among our men in Brazil. The body was interred in the forest near the bank of the Madeira and about two hundred miles above the mouth of that river.

Under the personal leadership of Mr. Thomas Collins, affairs at San Antonio at once took on a brighter aspect. Work was resumed with increased energy and determination. The erection of a number of buildings was commenced.

These included a residence for Mr. and Mrs. Collins, another for Mr. O. F. Nichols, the resident engineer; a large store, to be stocked later with general merchandise and supplies of all kinds likely to be required by laborers; houses for storage, hospital, and other purposes; a new sawmill; and a bakery with items sufficiently large to bake all the bread required. On a high knoll, commanding a view of the falls and of the river above and below, foundations were laid for the main office building of the railway. Unlike some of the others, this was intended to be a permanent structure.

It was designed to be the largest and most conspicuous building in the place, two stories high, with verandas ten feet in width, completely surrounding on both floors. The first story was intended for office purposes exclusively, but the second was to be divided into spacious apartments, as a place of residency for the principal employees of the railway in San Antonio.

With all this work in progress, the little American settlement soon presented an appearance of businesslike activity in marked contrast with the air of sleepy indolence that seemed to pervade more pretentious places on the river below.

The laborers brought down on the *City of Richmond* were decidedly inferior to those who had embarked on the *Mercedita*. The wreck of the *Metropolis* had served to check the enthusiasm of the better class of laborers for South American service, and the contractors were therefore compelled to accept the men gathered from the slums of several of our large Eastern cities. The Italians, particularly, were a worthless lot of vagabonds from the start, Several newspapers predicted trouble with them, even before the *Richmond* had passed the capes of the Delaware.

At Philadelphia, they were perfectly willing to work for almost any wages the contractors chose to offer, but when they arrived at San Antonio and discovered that they had agreed to accept much lower wages than the skilled American and Irish laborers who had preceded them, their dissatisfaction and indignation soon culminated in open revolt. On arrival, the Italians were assigned to excavating a portion of the road bed in San Antonio but almost immediately threw down their tools and demanded an increase of wages from a dollar and a half to two dollars a day and board while flourishing pick handles about their heads, as an argument. Mr. Collins personally ordered them to resume work. His usually vigorous manner of expressing himself on such occasions fully compensated for any lack of fluency in the Italian language and left no doubt upon the minds of his audience as to the result of a failure to comply with his wishes. The reasoning seemed, for a time at least, to be convincing, and the Italians resumed work forthwith.

During the next two days, they quietly accumulated a supply of weapons of every description—guns, revolvers, knives, dirks, bludgeons, machetes, and axes. At noon on the twenty-seventh, they raided a little house near the sawmill, seized all the provisions they could lay hands on, and openly threatened to capture and appropriate the entire stock of arms and provisions stored in the warehouses at San Antonio. Their numerical strength caused this to be regarded as more than an idle boast. They numbered 218 men, and there had been so much dissatisfaction among our other laborers, on account of the privations they had suffered and their inability to secure compensation for their services, that it was impossible to predict what accessions this force might receive from malcontents of other nationalities. Of the original engineer corps, about fifty-three in number, fully one-half were in camp far up the river, accompanied by some of our best laborers, serving as chainmen, cooks, and axmen. Many of those who came down on the *Mercedita* were out of reach, engaged in felling timber to supply the sawmill or clearing the line preparatory to grading. Consequently, the whole responsibility of guarding, day and night, the property of the contractors against an attack from this band of cutthroats devolved upon the engineers and few clerks, whose regular duties were sufficiently arduous without this additional imposition.

Fortunately, there were no courts and no lawyers to get out injunctions, no politicians anxious to acquire popularity by expressing sympathy with crime, no Boston detectives to precipitate bloodshed where disciplined coolness and firmness were required, and no frightened militia ready to open fire at their own shadows. At first, no attempt was made to interfere with the movements of the strikers. Mr. James T. Brown, chief clerk, organized a force to constantly guard all approaches to the warehouses. The strikers were refused further supplies of food, but no other measures would probably have been taken against them had it not been for the defiant attitude they continued to maintain and the renewed threats of vengeance they proposed to execute.

It soon became evident, too, that our small armed force could not long perform their legitimate duties and maintain a continuous watch over the widely scattered property in San Antonio.

Accordingly, at 6:00 a.m. on the twenty-eighth, Mr. Brown, at the head of about forty armed men and accompanied by Mr. Collins and Captain Stiles, made an unexpected descent on the Italian camp before they had eaten breakfast and in a few minutes completely surrounded them. The immediate surrender of ten or twelve ringleaders, who were named, was then demanded. The others were assured of continuous labor at stipulated rates, should the demand be complied with, but in the event of refusal, they were told that they would be confined to their camp and deprived of food and assistance from without.

This proposal was greeted with hooting, yells, and derisive laughter. Mr. Brown then selected ten men out of the party to accompany him, and instructing the remainder to shoot fast and long should any attempt be made to resist or escape, he entered the main house in the camp, handcuffed some six or eight of the ringleaders found there, and brought them without a blow or shot. Only one man escaped the dense growth of brush in the rear of the house. The prisoners were at first placed under guard in front of the warehouses, while a better place of confinement was being constructed.

This new jail was a model of its kind, as regards simplicity and security. On a piece of level ground near the steamboat landing, a layer of iron rails was laid in close contact, with the rail base alternately up and down, so that they fitted into each other like groove-and-tongue boards and formed an almost solid iron floor. On this layer of rails, a second was laid exactly similar to but at right angles with the first. Above this double floor, a cage was then erected of iron rails, laid like the timbers in a log house. The prisoners were then placed inside, and over their heads a double roof, exactly like the floor, was placed. It is needless to say that escape, without assistance from the outside, was an absolute impossibility, and it was possible for one armed man to take care of all who would be crowded within.

Those arrested were consigned by the next steamer to the care of the Brazilian authorities at Manáos, and as no one took the trouble to prosecute them, they were subsequently released and made their way to the United States, where their tales of the cruel treatment they had received were circulated by the newspapers from Maine to California. After removing the leaders of the mutiny, a deadline was drawn around the Italian camp, and sentinels, four at a time, regularly paced back and forth under the most positive instructions to shoot the first man who attempted to leave the place.

The strikers were deprived of all the provisions they had stolen, and to guard against a possibility of their overpowering the guards by a united onset, the signal gun of the steamer

Richmond, lying nearby, was loaded with buckshot and trained on the Italian camp.

After the events just discussed, Mr. George W. Creighton, now general superintendent of the mainline of the Pennsylvania Railroad, who had been on guard much of the previous night and assisted by making the arrests, returned to headquarters and endeavored to obtain a little sleep, but finding that impossible, after a time, he thrust his feet into a pair of native house slippers, destitute of heel covering, and sauntered down the path to the steamboat landing. As he approached the rear of the Italian camp, he observed one of the guards, about two hundred feet distant with his rifle out of reach on the ground, engaging in a desperate hand-to-hand struggle with the Italian who had escaped arrest that morning. Before Mr. Creighton could reach him, the Italian drew a small .22-caliber revolver, shot his opponent, and ran for the thick brush that everywhere bordered the beaten path, hotly pursued by Mr. Creighton, who had picked up the rifle as he passed and lost both slippers in the chase. The rifle was practically worthless as a means of shooting the fugitive because the dense brush made it impossible to point it readily in any given direction while the Italian was constantly flourishing a smaller but more serviceable weapon and threatening to fire. Fortunately, a native Indian, who happened to be in the way, instantly grasped the situation and seized and held the Italian until Mr. Creighton came up and, with the muzzle of his rifle close to the man's ear, marched him off to join the other prisoners.

The wounded man was then carried on board the *Richmond* and our three physicians summoned to attend him, while an anxious crowd gathered at the landing, to learn the probable result of the shooting. Up to this time, our medical practitioners had not been favored with many opportunities to demonstrate the skill and ability they unquestionably possessed, nor had they been at all conspicuous in the life at San Antonio.

The only dangerous case they had treated had terminated in the death of the patient. No one affected by the common complaint, malarial fever, ever thought it necessary to consult these gentlemen, unless his private supply of quinine was exhausted and he required a prescription to obtain it from the drugstore. Their practice had, therefore, been confined almost exclusively to the laborers in and near San Antonio.

Finding themselves suddenly thrust into the limelight of publicity and unexpectedly occupying the most prominent positions in the drama then being enacted, they at once assumed an air of dignity and importance befitting the occasion, the first dangerous surgical case that had occurred in their Brazilian practice. They commenced issuing bulletins regarding the condition of the patient, duly signed by each member of the mighty triumvirate, upon whose efforts the issue of life and death depended.

As the sympathetic crowd was not equipped with unabridged dictionaries to interpret the mysterious language in which medical men love to conceal their ideas, the effect of the announcements made from time to time was to increase rather than allay their fears. The plain English of the documents promulgated seemed to be that the place where the shot had taken effect and the direction of the bullet both appeared to indicate that the patient could by no possibility survive but that the doctors could not possibly state the case to be hopeless until they had probed for and extracted the bullet, which, in the extremely weak state of the patient, they could not safely undertake to do until later.

This sort of thing had continued for probably two hours when the victim, impressed with the idea that there was something essentially immoral in such a termination of existence, begged the doctors that he might not be permitted to die with his boots on. The objectionable articles were then removed, and out of the leg of one boot, the bullet fell on the floor.

It had only inflicted a flesh wound. The disgusted disciples of Æsculapius gave the wounded man a double dose of whiskey to steady his nerves after the fright into which their ominous looks and sesquipedalian words had driven him, and the dying man was soon walking about among his friends on shore, evidently convinced that, however inappropriate for perambulating the golden streets of the New Jerusalem, his long boots could not be discarded while he continued to frequent the tropical jungles of Brazil.

The arrest of the leaders and the want of food soon brought the Italians to terms, and on the morning of March 30, they sent word to Mr. Collins that if he would give them something to eat, they would gladly resume work. Mr. Collins replied that he had already fed them all the way from Philadelphia to San Antonio without any return and that he did not propose to vie them another morsel to eat until they had done an honest day's work. They had to surrender on his terms and were immediately put to work. No similar trouble occurred afterward.

As the writer now attempts to recall the life at San Antonio, assisted by brief notes made at the time, he finds that in regard to the unpleasant features of existence there—the monotonous round of daily labor performed under most trying circumstances, the privations and discomforts incident to work on the line, and the experience with sickness and death—his memory often proves defective, while little incidental occurrences of an amusing character, really very few and far between, loom up from the distant past with a vividness truly astonishing and constantly seek a more prominent place in this narrative than their importance deserves. One of these incidents occurred shortly after the arrival of the *City of Richmond*.

One of our engineers, who had taken a long Sunday afternoon tramp through the forest for the purpose of increasing this collection of stuffed birds, returned to headquarters pale and breathless. He reported having encountered four large jaguars on the banks of the stream that empties into the Madeira above San Antonio. He was armed only with a shotgun, and his cartridges contained nothing heavier than birdshot, so he deemed it advisable to beat a hasty retreat, followed for some distance by his newly made acquaintances. Among those who head the story were two men in whose minds some evil genius inspired the thought that it would be a good thing to steal a march on their companions and achieve immortal fame by being the first to kill one of these dangerous animals.

Both of them had been brought up in the populous districts of our Eastern states, and, up to that time, their only knowledge of such animals and their habits had been obtained from zoological gardens and the books of adventure usually supplied to schoolboys in order to gratify their omnivorous and indiscriminate appetite for tales of life among savage men and beasts.

Accordingly, at the close of office hours on the following day, provided with one Winchester rifle and two .38-caliber revolvers, some eatables, matches, tobacco, a hatchet, and one large rubber blanket, they started over a path made by wild animals along the north bank of the stream before mentioned and at the end of an hour reached the locality where the jaguars had been seen the day before.

As darkness was rapidly approaching, they at once gathered a supply of comparatively dry wood, kindled a fire, ate their lunch, and prepared for an all-night interview with the denizens of the forest. The dense foliage made the place gloomy even at midday, but when the sun set, the light of the fire only seemed to penetrate for a few feet into the forest, and everything beyond was enveloped in the most complete darkness that can possibly be imagined.

These arrangements had scarcely been completed when rain began to fall in torrents and continued to pour down throughout the night.

Our Nimrods soon found their fire extinguished, their fuel too wet to burn, their matches ruined, their tobacco soaked, and the all-pervading darkness so impenetrable that one could not ascertain the position of the other except by voice or touch. They stretched themselves on the ground covered by the rubber blanket, in every depression of which a puddle of water formed, but their repose was of short duration.

Out of the depth of the forest came unearthly and, to

them, unfamiliar sounds. Birds screeched, monkeys howled, and other animals growled and snarled, while every now and then, a large branch would crack and break under the tread of some beast that they knew must have been very heavy to produce the resulting noise.

Frequently, these sounds originated at points quite near to the jaguar hunters. Then up would go the rubber blanket, down their backs would flow the water that had accumulated in its depressions, shots would be fired, and for a few minutes, quiet would prevail. There was great danger that in case the men were molested by wild animals, they might become separated and one shoot the other. They, therefore, endeavored to lie in close contact, back to back, so that each one would always shoot in a direction opposite to that of his companion. At last, one of the men, whose defective hearing rather than his courage rendered him less sensitive to the pandemonium around him, fell asleep only to be awakened a few minutes later by the other kicking his shins and violently remonstrating at the failure to perform his full share of duty in the common defense.

With a yawn, the half-awakened sleeper inquired, "What is the matter?"

His companion answered, "There is a wild beast directly in front of me, not twenty feet away, and the large branches break under his feet as though he carried the weight of an elephant."

"Well, shoot him then! You could surely hit an elephant at twenty feet," slowly drawled the deaf and sleepy individual.

Another series of violent kicks followed, accompanied by the exclamation, "Great heavens, man! There is another on your side, and you are not half awake!" Both then opened fire in opposite directions, and a few minutes later, quiet was again restored. The rain continued to fall as it could only in the tropics. The clothing of the men was as wet as it could be and the ground upon which they were lying was simply a mud puddle, but the blanket was still used as a covering most of the time in order to keep the firearms as dry as possible. The one nap previously described was the only one permitted that night. At every failure to respond to a question or remark, the kicking process was renewed with such vigor that to protect himself for the suspicion of being asleep, the person who had previously offended in that respect began to quote:

> How dainty sweet it were reclined
> Beneath the wide outstretching branches high,
> Of some old wood, in careless sort to lie,
> Nor of the busier scenes we left behind aught envying.

This was regarded by the unappreciative individual to whom it was addressed as an indication of undue levity on the part of the speaker and called forth an earnest remonstrance, accompanied by sundry inelegant expressions and intimations that a man who could repose all night in mud and water surrounded by savage beasts and think of such inappropriate nonsense was mentally unfitted for the serious business of jaguar hunting.

All through the night, one alarm followed another and so much ammunition was expended that it became doubtful whether the supply would last until morning. The hunters fully expected to find dead animals piled around them, but strange to say, when daylight came, not one could be found in the vicinity, and they hastened back to San Antonio, there to endure the merciless chaffing of their comrades, who manifested a rare talent for adorning the tale with many additional attractions, which originated entirely in their own fertile imaginations.

> All who told it added something new,
> And all who heard it made enlargement too;
> In every ear it spread, on every tongue it grew.

While the drastic methods employed to suppress violence and disorder had forced the Italians into good behavior, there were some hot-bloods among them who yielded only a sullen and reluctant submission. From time to time, rumors reached headquarters that they would yet attempt in some way to get even with "the padrone," as they called Mr. Thomas Collins. Occasionally, reports of intended desertion by these men took on a more or less definite form, but as they had no boats or money to pay steamer passage down the river and as every way of escape overland was impracticable, no one paid any serious attention to these threats until one morning seventy-five or more of them were found to be missing. Investigation revealed the fact that they had set out for Bolivia overland through the dense and all-but-impenetrable forests without map, compass, or provisions. Nothing was ever heard of them afterward, and there can be no doubt that they were lost in the forest, perished of starvation, or, worse than that, themselves served as food to gratify the not-too-dainty appetites of the anthropophagus Parentintins, a tribe of savages whose existence we were all, at that time, inclined to doubt in spite of many tales current in regard to their depredations.

As I am illustrating what absurd things men of experience and good sense will do under conditions with which they are unfamiliar, it is appropriate here to mention in its chronological order an incident in itself of trivial importance.

Mr. Collins had observed on his way up the Amazon the method of supplying live cattle for food on the steamers. He saw that these cattle were all of the well-known Durham and Jersey breeds and probably supposed them to have the same characteristics as the gentle, docile animals of the same ancestry at home. He was not aware of the fact that under their new environment—roughly treated; poorly fed; plagued by insects, reptiles, and wild beasts—the descendants of those originally imported to Brazil became as wild as any denizens of the forest and some of them positively dangerous. Mr. Collins purchased a number of these cattle and had them transported by steamer to San Antonio and turned loose where there were no fences to limit their choice of a place of residence to any particular part of the South American continent. He doubtless expected them to graze at will during the day, confident that they would return at milking time and each patiently await his or her turn to be slaughtered, as the demand for fresh beef might require.

The cattle did not take Mr. Collins's view of this situation; neither did they equip themselves with the proverbial hay on their horns, because there was no hay, but after forcing all the residents of San Antonio to take refuge in buildings, they made for the forest and, like the Italians, were never seen again. Certain it is that the writer cannot recall having tasted a single morsel of fresh beef during his entire residence at San Antonio and that he *does* recollect most distinctly, on several Sundays thereafter, unsuccessfully hunting for "Tom Collins's wild cows" with a rifle, not merely for sport, but to satisfy his cravings for fresh beef and incidentally to earn the officially promulgated reward of five dollars for the capture of each animal dead or alive.

XVI. With Byers at Macacos

*His heart was in his work, and the heart
giveth grace unto every art.*
—Longfellow

Of the four engineer corps in the field, it is simple justice to state, the one under Mr. Joseph Byers was preeminent as the first to leave San Antonio and plunge into the unknown wilderness above, as the one that accomplished more actual work than any other, and as the last to yield to circumstances beyond their control and abandon the enterprise. The writer greatly regrets, therefore, his inability to find data upon which to base a continuous history of its progress and achievements. Mr. Byers has long since gone to his eternal campground, but while a man exists who knew him on the Pacific railways of the West, in the forests of Brazil, on the railways of Mexico, or while engaged upon public works in the Republic of Colombia, no monument or history will be required to keep his memory green or to maintain the respect and affection with which he was regarded by all who knew him.

When quite a young man, he enlisted in the United States'

regular cavalry and had the usual exciting experiences while fighting Indians in the West. He was taking cypress timber out of the swamps of Louisiana when the Civil War broke out and promptly abandoned that occupation to become a captain of infantry in the Confederate service. After the death of his commander, General Albert Sidney Johnston, he resigned and soon after joined the light artillery under Major Pelham. During the war, he was twice captured by the Union troops and at all times bore the reputation of a fine gunner and a gallant soldier. He was too modest to refer to his military career and few of us ever heard of it until after our return from Brazil. In association with his equals, he never resented being spoken to as plain "Joe" Byers, and "Mr. Byers" was all he exacted from his subordinates. He belonged to that class of railway engineers to whom this country is almost exclusively indebted for the tremendous railway development that culminated in 1873. He was a typical engineer of the period when young graduates of engineering schools were reluctant to confess the possession of a diploma and when familiarity with fighting, firearms, fire water, and faro was not regarded as an unnecessary finishing touch to complete the training of the thoroughly accomplished and "practical" railway engineer. Warm-hearted, fearless, and indefatigable, he bore with a grim smile of stoic indifference every misfortune and looked with supreme contempt upon those who grumbled at semistarvation or murmured under attacks of the prevalent fevers.

Two letters bearing dates of March 20 and 25, written from Camp Macacos, five miles above San Antonio, by Mr. Charles J. Hayden, then attached to Mr. Byers's corps, but now right-of-way and tax commissioner on the Great Northern Railway, afford a glimpse of camp life at Macacos that is interesting. From these letters, the following extracts are taken:

> Our camp is on the river bank and commands a fine view both up and down the Madeira. Mr. Runk's party is camped at the same place ... The country here is an almost impenetrable forest. The vines are the greatest obstruction to us in working. They vary in size from a diameter of three or four inches down to fine threads and are as tough as leather. You will understand how hard it is to cut through the forests when I tell you that with five axemen the greatest run our party ever made in one day was 3900 feet, while in the United States we frequently ran four or five miles over rough country ... People at home grow enthusiastic over the luxuriant forests of the tropics but when you get into them here and find you can't see ten feet from you in any direction, your enthusiasm soon vanishes ... We have been disappointed in the number of wild animals and snakes, but there is one pest we did not count on that gives us no end of trouble and annoyance. I refer to the ants.
>
> They are of several varieties that vary in size from one and a half inches in length down to proportions almost microscopic. They swarm over the ground, on the vines and trees, get into our victuals, our clothing and, in fact, into everything. They possess the peculiarity common to nearly all insects in this country; they bite. There is one species, fortunately scarce, whose bite is said to be deadly poison. They

are about an inch and a quarter long, have a velvety appearance and an enormous head. Another variety, much smaller, but far more numerous, worry us more than all others combined. They are red in color and in the act of biting, double themselves up so that the ends of their bodies meet. The bite causes no swelling but feels to the victim like an electric shock penetrating to every part of the body. On our line there is a wide creek, twenty feet deep, which we have to cross on a fallen tree. The trunk is rather slippery and some skill is required to cross on it, even under favorable circumstances. The ants have now taken possession of this rude bridge and every time we attempt to cross their biting causes some one to lose his balance and fall into the creek. The bite of the large black ant (one and a half inches long) is very painful and causes a terrible swelling ... Our domestic pets are lizards and some of the prettiest colors are blended on their backs—brown, green, white, black and red. They become quite tame and run about camp as though they belonged there. The only objection to them is that they get into our hammocks and, when we get in and put our feet down under the blanket, the cold slimy feeling of their sleek bodies sends a chill through us.

Another pet, or rather an animal that aspires to be a pet, is the dreaded tarantula. These are the most horribly ugly animals we have seen yet, as large as the palm of your hand, black and hairy. We take good care to shake our clothes and examine our boots closely before putting them on in the morning, particularly since Brisbin found one in his boot. Monkeys are plentiful. We have had monkey roast several times. The meat tastes very much like rabbit or squirrel. Our greatest luxury is tapir. The flesh of this animal is just like tender, juicy beef, a most agreeable substitute for mess port and beef, when in rare instances it can be procured. At times we were treated to turtle soup. The turtles are enormous, some measuring four or five feet across. We have heard many turkeys, but, so far, have been unable to shoot any ... The character of the ground is rugged and rocky for several miles, as might be expected from the fact of there being such a series of falls and rapids in the river itself ... In running our cross lines we traversed ground upon which no white man had ever trodden before and really I cannot see why any respectable white man should desire to be here. It is a perfect wilderness of vines climbing over palms and giant trees of other kinds, cut up with deep, narrow gorges having rocky precipitous sides, at the bottom of which flow streams clear as crystal.

When the *City of Richmond* arrived, we at Camp Macacos had been grumbling a good deal at having only heavy corn bread and weak tea to live on. Our meat was all gone. We have no lard or grease to bake cakes with. We had sent in one requisition after another to headquarters and all had been quietly ignored. We felt desperate ... Yesterday was an eventful day. We had two and a half miles to walk to work over the roughest kind of country, crossing six ravines, all over 100 feet deep. The line then ran down the bed of a stream into a terribly wild and tangled network of vines and reeds. The atmosphere was hot and stifling. We were working in mud and water over our boot-tops all day. Toward evening, just as Bruce was thinking of stopping for the night, we came to an immense boa constrictor, which lay coiled upon the branch of a tree near the stream. He showed fight immediately. We had a rifle and several revolvers in the party, with which we opened fire on him. The first rifle shot penetrated his neck just back of his head and several pistol balls struck his body. Still he was not killed, but kept advancing rapidly, we retreating at the same time. Another volley stopped his progress, this time the rifle ball entering his head. He continued to squirm and wriggle and it was terrible to watch his contortions. There was a large branch lying there, around which he wound himself and broke it all to pieces. Volley after volley was poured into him from the revolvers until he became rather quiet. Then the men attacked him with axes and brush hooks, which soon finished his career. It was the largest snake I ever saw anywhere, measuring twenty-five feet in length and ten inches through the body.

We then started back to camp, straggling as we always do on the return trip, some hurrying ahead, others, for one reason or another, lingering in the rear. This time one of the party, named Delleker, got behind and in attempting to take a short cut lost his way. We did not miss him until we reached camp, even then supposing him to be close behind us. It was after dark and we had supper as soon as we could get ready for it. Still Delleker did not appear. Then we began to feel uneasy. A party with lanterns started to find him. They went more than a mile firing pistols and shouting without hearing any response. It rained in torrents all night and this morning another party started out as soon as it was daylight. They found Delleker about two miles from camp on another line, wet, frightened nearly to death and worn out with hunger and fatigue. He had passed a fearful night … Four men in a boat, stolen from the steamship *Mercedita*, attempted to go up the river with their baggage and provisions. When they came opposite our camp this morning, some of our men saw them, and knowing of the disturbance (made by Italians at San Antonio) hailed them. At first, they would not approach us, but, on being covered by half a dozen rifles, concluded it was better to comply with our orders. They did not know where they were going and told very conflicting stories. Mr. Byers sent boat and men back to San Antonio under guard.

XVII. A Preliminary Reconnaissance on the Upper Madeira

Men of action these—
Who seeing just as little as you please
Yet turn that little to account.
— Browning

On April 3, 1878, two parties left San Antonio, bound for points on the Upper Madeira, their routes coinciding as far as San Carlos. The first included Corps No. 4, under Captain Amos Stiles, and the second consisted of Mr. Othniel F. Nichols, the resident engineer and official representative of the Madeira and Mamoré Railway Company; Mr. Charles M. Bird, chief engineer for the contractors; Mr. Charles W. Buchholz, the principal assistant engineer in general charge of surveys; and Don Ignacio Arauz. The last and smaller of the two parties contemplated making a reconnaissance with a view to determining in a general way the best route for the railway and deciding the important question of whether it was better to follow the riverbank closely or to run our lines some distance inland. From a very interesting paper prepared by Mr. Nichols, the writer is permitted to quote the passages relating to this reconnaissance, which occupy the remainder of the present chapter.

The start from the landing above the Falls of San Antonio was energetic if not impressive. The small steam launch, barely large enough to hold comfortably the four white passengers, carried, in addition, four Bolivian Indians employed by Sr. Arauz and towed two of his canoes, laden with an extra supply of coal.

The river current was strong and the launch did not steam well, so the canoes were finally set adrift to make the best progress they could with the Indian paddles.

Relieved of its heavy tow the launch soon reached the Falls of Theotonio, where it was taken out of the water and dragged over the rough portage by twenty Indians taken from one of our own canoes and those which had carried Captain Stiles's corps.

The camp of our combined parties was established at San Carlos, between Rosstown and Morrinhos, on April 6th. There an independent wall tent accommodated Bird, Buchholz, Arauz and myself, while Arauz's Indians cooked and served our meals and handled our stores. The establishment of Stiles's camp seemed to arouse the ants, and several of the men were badly bitten, some of them by a large black ant nearly an inch in length, resembling a wasp without wings and with fully as great stinging capacity.

At noon on the 8th, we reached Don Pastor Oyola's residence at Concepcion del Morrinhos, where he lived the life of a typical Bolivian *Seringueire* (rubber gatherer) of the Madeira. Oyola had some twenty or more acres of land cleared about his large house and had cleared the island of the opposite, or eastern, side of the river, where he expected to pasture cattle. The dining table was spread under an immense canopy of netting to keep out the mosquitoes, which seemed more abundant than elsewhere on the river.

Of the Bolivians we met on the Madeira Oyola and Arauz represented two distinct types. Oyola was a business man, a man of energy and executive ability. He never used a hammock, had a bed in which he slept—and sat in a chair during the day time. His buildings, store-house and canoes all indicated the man of action and of thrift. He had seringals above and below his residence, employed about fifty or sixty servants on his place, spoke Spanish only, but wished he could speak English. He was a large, heavy, but active man and perhaps thirty-five years old. Arauz, on the other hand, was a smaller man, thin, wiry and probably about forty-five years of age. He boasted of his hatred for work and claimed to be "a drone or queen bee as you will." Slender, with finely cut features and of marked intelligence, a man of the type of Machiavelli, he understood English well, or at least we thought he did, and yet never spoke anything but Spanish. He had been a careful reader and was a good writer.

Colonel Church met him at La Paz in 1868 and

he wrote an essay at that time on "The Fluvial Outlet for Bolivia." He was, however, more of a dreamer or schemer than Oyola and a man likely to exercise far greater influence and power in a political intrigue. His conversation was always to be critical and often cynical and sarcastic.

Oyola accompanied us up the Jaci Paraná, a stream that entered the Madeira seventeen miles below the Caldeirão do Inferno, and then returned to his home, while we continued up the latter river.

In the early evening of April 10th we tried to shoot some ducks from the launch, but were unsuccessful, partly because our guns were loaded with bird shot only. We could not get Arauz to shoot. He laughed at our failure and finally said, "I know where I can get a duck—in my barn yard," and he would not stop for dinner but pushed on until we reached the Caldeirão where he served his domestic duck at nearly 11 P.M.

Arauz's establishment at the Caldeirão was beautifully located on high ground near the river and within hearing of the incessant roar of the falls, to which we soon became accustomed. Once settled in his house we entered with zest into the spirit of the *seringueiro's* life.

The abodes of the well-to-do *seringueiros* on the Madeira are all of the same general type, a two-story house with frame of heavy timber, the lower floor literally on the ground, the ends enclosed to form store-rooms, the middle left open with a broad stairway leading to the second story, where the dining and sleeping rooms are located, and outside of them a broad veranda extends, often completely around the building. Frequently the second floor has a central apartment opening on the veranda and intended for use as a living and dining room. With this the end rooms, generally reserved for private or sleeping apartments, are connected by doors. There are no windows and generally no ceiling. The sides, partitions and flooring are made of slabs of the black palm about four inches in width. The soft interior of the palm works easily to a flat bearing surface, which is placed inside for the siding and on the underside for the flooring. The hard, round, gray exterior of the palm slab gives a rather pleasant external appearance to the building, and produces a rough but durable floor. All the doors, casings and similar fittings are made of the soft pale-red cedar, which floats down the river and becomes thoroughly water seasoned during its journey from the upper Madeira or the River Beni. It is easy to work and never attacked by ants, which eat up all soft woods found in the vicinity.

These *seringueiros* lead the life of wealthy farmers or country gentlemen, surrounded, like mediaeval barons with a retinue of Bolivian servants and their families. They have credit in Pará, in return for rubber shipped there by steamer from San Antonio, and their agents are always ready to send them any articles of comfort or luxury which they desire. Oyola had quantities of silks, silk umbrellas, Panama hats, etc., etc.

This baronial life is a glorious one in some respects.

These men are absolute masters of their peons. They have the choicest of native foods, game from the forest and fish from the rivers. They raise cattle and, to a limited extent, plant crops, but only for use at their own establishments. They travel away from their estates entirely by canoe on the rivers and these canoe trips, of ten or fifty miles or even far into Bolivia, are such no doubt, as the pioneers had in the early days of American development on the Ohio and Mississippi rivers and their tributaries. Here, however, the tractable employees and the warm climate make the experience generally more enjoyable.

Arauz, Oyola and Mercado really controlled trade on the upper Madeira River for more than two hundred miles. These men we learned to know and respect and we were indebted to them for royal hospitality and for assistance on many occasions.

Their manner of life showed that the wonderful productiveness of this region could be utilized for the support and comfort of myriads of men and impressed forcibly upon us all the futility and fatality of any attempts to ignore the experience of this baronial life by those engaged in the conduct of important engineering and commercial enterprises in the valley of the Madeira.

At the Caldeirão Arauz had a crude sugar mill with wooden rollers worked by oxen. After extracting the cane juice, it was boiled in flat cooper pans over a brick fire place. He made no attempt to refine the product, and produced only crude molasses.

We knew Arauz had his wife with him at the Caldeirão do Inferno, because her woman's hand was evident in all the little details of our entertainment, but we never saw her. Mercado's wife was with him at Paraiso and San Antonio, where he built a town house. Oyola's wife was absent on a long and tedious journey to Bolivia in quest of peons and to visit friends.

Bird left us at the Caldeirão and returned to San Antonio in the team launch, leaving Buchholz and I to continue the trip up the river. It was, however, nearly impossible for us to get away. Arauz was not ready to go, did not seem at all himself and continually postponed the start. We were his guests and could hardly do more than hint at departure. He was drinking heavily of "Three Star Hennesy," a terrible stimulant then almost universally used in the tropics, much purer than the crude native rum, distilled from sugar cane and known as *cachaca*, but even more deadly in its effects, especially upon the Anglo-Saxon. We made many vigorous efforts to get away until one morning, about a week after our arrival, Buchholz said, "We shall start to-day. Arauz is on his last bottle of brandy and, when that is gone, we shall go." Just then Arauz appeared at the door of his room, which closed quietly after him, with a bottle of brandy in each hand and one under each arm. "Where did you get those?" said Buchholz. *"Tengo una mina abajo"* (I have a mine below), replied Arauz with that cynical, and only too familiar, smile of his. We always believed that he had heard and fully understood Buchholz's remark about our departure.

The joke, nevertheless, seemed to delight and revive the whimsical Bolivian and we did start soon after noon, entering a large and commodious canoe, which had been drawn from the lower level to a point above the falls on the previous day. This canoe was about thirty feet long by six feet wide. The bottom was formed from a tree trunk hollowed out and dressed down to a thickness of two or three inches. The sides were made by spiking a board to this bottom on each side and a second or even a third to those preceding; so that the finished canoe had the appearance of a clinker-built boat, broad and flat in the middle and roughly pointed at the ends. We were on a passenger and not a freight trip; so carried only provisions and camp equipage piled in the middle of the canoe. Six men paddled, three on each side seated close to the edge, while a seventh man stood in the stern to steer. The *camarote*, a small arched framework, thatched with palms and about six feet long, covered the canoe near the stern and under this shelter three or four persons could either sit or recline on the rough quarter deck.

The stroke of the paddles was constant and effective, the men making about forty strokes a minute, while the helmsman or captain, tapped the time with his heel to encourage them at particularly bad places in the river.

It was singularly pleasing to sail by moonlight on this great tropical stream. The air was cool, the men worked better and the canoe would glide along now on the broad open river, again, in swift currents near the banks and under rich tropical foliage, dense to the water's edge, where the silence was broken only by the sound of paddles striking the water, the howling of monkeys and shrill screeches of night birds on shore.

The first night after leaving the Caldeirão we camped not far above the rapids leading to the falls, erecting our tent over a framework made from trunks of small trees. In addition to the usual tent poles, four posts, with crotches at their upper ends, were set firmly in the ground inside and at each corner of the tent. Upon these were fastened four stout horizontal poles, which, together, formed a quadrangle around the interior and furnished support for our hammocks.

We generally kept our outer clothing off the ground, sometimes by winding it into our hammock strings. At day break when Buchholz reached for his watch, which he had tied to a tent pole by its silken cord, it fell into his hand at the first touch. The ants had severed all but a single strand of the silk. The inner lining of a straw hat, which had fallen to the ground, had been cut out entirely and carried away, while the softer upper leather of Arauz's shoes and the sole-leather covering of a rifle case had been cut into shreds. It seemed as if a thousand ants had been busy while we slept, but only a few red ones could be found about the tent in the morning.

These ants were about half an inch long, with large heads and the leather showed clean crescent-shaped cuts, as if the ant had stood on its hind legs and described the arc of a circle with its head while cutting the leather.

Our tents stood at the river's edge and about six feet above the water. Behind us the ground rose abruptly out of a species of canebrake, which nearly surrounded the small clearing we occupied. On the higher ground stood great trees, from which immense vines hung in picturesque festoons. The moon shone brightly all night and the howling of monkeys and shrieks of night birds, varied occasionally by growls and roars of larger animals, made us realize that we were not entirely alone in the forest. The night air was surcharged with moisture and, on awakening in the morning, every article of apparel, not worn on the person, was usually quite damp.

We reached the Falls of Girão before noon and the view approaching them was beautiful. Isolated peaks, covered with vegetation, appeared beyond the dense forests on the river bank, which, though not very high, served to relieve the monotonous uniformity of the tree bounded picture.

We landed in a little bay well shaded by large trees and, leaning against one of their trunks, found a few lances or harpoon shafts.

Walking around the eastern end of the falls, over the rough pathway by which canoes were hauled from one level to the other, we breakfasted at the upper landing place and then cut our way in an easterly direction into the forest for a distance of a mile or more, finding the ground quite hilly and rising possibly one hundred and fifty feet in the distance traveled. It rained hard during the trip and we got thoroughly wet. The trees were less frequent, not so large as those near the river and were of harder and more durable wood.

At this higher elevation there were very few vines on the trees and but little under-growth save palms, whose fronds, rising directly from the ground, extended in graceful curves for their entire length of from fifteen to thirty feet and furnished the material used by the natives in thatching the roofs of their houses. Laid with a steep pitch, such roofs are impervious to rain and sun, but permit the air to circulate freely through the interstices.

Resuming our voyage, we landed at a seringal (rubber gatherer's house), where we passed the night, started again early in the morning and reached the establishment of Sr. Arauz at Tres Irmãos about 3 P.M.

Arauz had about thirty people at Tres Irmãos, mainly Bolivian rubber gathers and house servants with four Caripuna Indians. The two male and two female Caripunas remained at Tres Irmãos when others of their tribe left, driven out by the vanguard of civilization engaged in the quest for rubber. They lived by themselves in an old hut, fished a little and were even more indolent than the Bolivians. Simon, the elder of the male Caripunas, had to some extent adopted Bolivian ways. He wore trousers as well as a shirt and understood Spanish. The other was a genuine Caripuna of stocky build, almost sullen looking and not unlike Keller's pictorial representations.

Maria, the younger of the two women, appeared to be about forty, possibly not more than thirty, years of age; Isabel, the elder, was at least fifty and looked like a resurrected mummy from Peru. They all had the lobes of their ears punctured with holes, about three-eighths of an inch in diameter, to receive the tusks of the wild boar. So poverty stricken were these Indians that they had only two tusks among them and the punctures which they did not occupy were filled by short pieces of wood. The women had the partition wall of the nostrils pierced to receive feathers. Maria wore two short macaw feathers in her nose, giving to her the appearance of a person with a short stubby yellow and blue mustache. Her forehead seemed to have had a single application of vermilion oil paint badly put on and she had red crosses of the same material on her cheeks. She wore a bright colored chintz dress, which hung loosely from the shoulders and was much soiled and bedraggled, while Isabel's sole garment was a white cotton shirt, store-made, of the usual size and shape.

In order to interest the Indians and give us an opportunity to see them, Don Augustino, the manager of the estate, offered them some liquor, from a bottle. Simon took the first drink and Isabel the second, both making terrible grimaces over the dose. The odor attracted our attention and we found that Don Augustino had made a mistake and treated them to kerosene instead of *cachaca*. Arauz assured us that kerosene would not hurt the Indians, but Don Augustino thought it prudent to give them a dose of olive oil and the compulsory administration of this antidote frightened the victims. The women, especially, became greatly alarmed and the expression of terror on their painted faces almost gave us hysterics. We slept well that night under mosquito bars made of a stout fabric, which protected us from large insects as well as from mosquitoes.

Arauz had several acres of land under cultivation at Tres Irmãos, some planted with corn, which was left on the stalk to dry, about half an acre with rice, the remainder with yuccas, bananas, etc. The principal work, however, was the collecting and curing of rubber. It was still too early in the season for this work, as the water in the river had not fallen enough to permit the *seringueiros* to reach the rubber trees over the *estradas*, or paths leading to them.

Among the trees of the Brazilian forest none was more peculiar than a great soft wood tree, frequently found near the river bank. The bark is smooth and of light gray color as in some of our beeches, but the trunk, at say twenty feet from the ground, was often from four to six feet in diameter. Nearer the ground great ribs projected radially from the trunk, forming giant braces to steady the tree. These ribs are from eight to twelve inches thick and often extend five or six feet from the trunk; so that a horizontal section just above the surface of the ground would be somewhat star shaped. One of these trees at Tres Irmãos, left standing behind a rice field, had a small scraggy top and only one limb, an immense one extending horizontally over the field at a height of about sixty feet from the ground. The spread of the tree at the ground was about twenty feet, including the braces. The diameter of the trunk, above the braces and about thirty feet from the ground, was about six feet. The trunk then tapered gradually and uniformly, like an unfluted column, for about forty feet to a diameter of perhaps five feet. There the great horizontal limb originated and extended over hundred and ten feet over the rice field. Parasitic vines covered the tree and limb, both of which seemed dead or nearly so. A number of vines hung vertically from the limb at intervals along its entire length; so that, at a distance, the tree looked like a great derrick with a huge horizontal boom from which loose ropes had been left hanging.

On April 17 we went up the Tres Irmãos River for several miles in a canoe and then walked over low lying land through a dense forest in an effort to find clear land which the Indians say exists to the south and east of the falls. We found nothing, however, but the same almost impenetrable jungle that covers the ground near the Madeira, on the higher lands more hard wood trees and less undergrowth, but always the same virgin forest.

It rained heavily on the return trip to Tres Irmãos, but we had raincoats and our canoe man had a cloth bag rendered perfectly waterproof by treatment with rubber. He folded his shirt carefully and put it in the bag as soon as the rain commenced and when it was over, half an hour later, he had a dry shirt to put on, which he wore while his trousers were drying.[18]

All the way from the Caldeirão do Inferno to Tres Irmãos hills were visible to the south and east at a probable distance of two or three miles. Some of these were detached and reached a height of five or six hundred feet, but were neither as high nor as picturesque as those on the opposite side of the river. They indicated that the high table-land below the Jaci Paraná did not continue south of that stream that these hills would force the railway line nearer the river from the Caldeirão do Inferno to Tres Irmãos; that good ground for the railway would be found along the edge of this hilly country and beyond the low ground and jungles on the river bank; and that this line would doubtless be shorter and better than if located near the river.

We spent nearly a week on the return trip to San Antonio, stopped at Don Pastor Oyola's again over night and cut through the jungle at two or three points,

---

[18] "From there we ascended the tributary, called also Tres Irmãos to look for the probable crossing of the proposed railroad; but we found all the adjoining country, except immediately at the mouth of the river, submerged by the flood—the water being then at its highest. We went in vain through the dense forest in a small canoe, to hunt for and see the great plain without trees, said to exist and extend all the way across the peninsula formed by the large bend in the Madeira River at this point. We waded in vain through swamps and marched over narrow, low strips of territory; we had not force enough, or time enough to reach the promised land this time. Besides, our Indian guide lost his way, and we regained the river with great difficulty during a heavy rain storm," from a letter of Charles W. Buchholz to Franklin B. Gowen.

reaching higher, drier and clearer ground and finding nothing at all discouraging for the construction of our railway. We felt very well repaid for our trip by the advance information obtained in regard to the characteristics of the middle region of country through which the line was projected, information that would have been invaluable had the railroad been continued to completion.

We felt that the character of the country and the physical difficulties to be surmounted had not been misrepresented, but that a railway could be built through it with comparative ease and within the estimates made therefore, if the proper amount of energy and attention could be given to preparation for the work and to care of workmen.

As I have had occasion to speak quite frequently of two members of the triumvirate which monopolized the rubber trade for many miles on the Upper Madeira, it may be of interest to mention here, a visit subsequently made to the home of the third member, Don Santos Mercado, though he lived in territory far removed from the scene of our railway operations. Mercado's plantation was situated at Paraiso on the east bank of the Madeira, 178 miles below San Antonio. We went down in a large canoe with six paddles, three on each side. Between the paddlers the baggage, camp equipage and other freight was piled. The party of four persons, two ladies and two gentlemen, occupied the palm covered shelter, or *camarote*, at the stern of the canoe, while a man with a long paddle, standing in the stern did the steering. On a still moonlight night the canoe drifted down stream unaided by the paddles, the Indians sleeping at their posts. The passage was made for miles down the middle of the river, where nothing disturbed the calm of a tropical night except the sound of an occasional splash at the stern, when the helmsman dropped overboard for a swim.

The house of Don Santos was of the type usually found in the Madeira Valley. The floors of split palm presenting a rounded upper surface, were as difficult to walk over as a corduroy road. The rooms were very meagerly furnished and contained no beds. I suppose we took the first modern bedstead up the river. There, as elsewhere, on the Amazon and Madeira, gorgeous hammocks were used to sleep in at night and sit or lounge in during the day. There were few chairs. Rude benches surrounded the dining table and the principal attraction was an immense Parisian music box that cost Don Santos upwards of one thousand dollars. He had immense fields of plantain and banana trees, spaced thirty or forty feet apart, with rows of coffee plants between.

Though the plantain belongs to the same family as the banana, is a stable article of food throughout tropical South America and, on the Upper Madeira, reaches an immense size, it is not sweet and is only eaten by human beings when cooked and served as a vegetable.

One morning a veritable mountain of plantains containing probably twenty or thirty large cart loads, was placed in front of the door. The cows were first turned in to feed on them and afterwards a herd of swine; so that few remained at the close of the day.

During our stay of three of four days at Paraiso, Don Santos gave a banquet, which had somewhat the appearance of a barbaric feast and was probably intended to astonish as well as regale his visitors. In the center of a large table was set a head of beef, with the horns still on, excellently cooked and served. The horns were utilized to steady the piece, the under jaw had been removed, the tongue was in place at the top of the dish and the whole was garnished with ears of green corn, plantains and other vegetables.

XVIII. Events at San Antonio Resumed

Lord safe us, life's an unco thing!
Simmer an' Winter, Yule an' Sprin,
The damned, dour-heartit seasons bring
A feck o' trouble.
—Robert Louis Stevenson

As noted in the previous chapter, after long delay, caused by lack of camp equipage and provisions, Corps No. 4, under Captain Stiles, finally started up the river to begin surveys at Morrinhos, on April 3, 1878.

Five days later, Mr. Collins began laying rails on the main line, and on April 9, he entered into a contract with Benjamin Huff to clear two miles of the roadway of timber and brush. The clearing was to have a uniform width of one hundred feet and be paid for at the rate of seven hundred dollars per mile. Any felled timber, suitable for the purpose, was to be cut into crossties, and for these, Huff was to receive twenty cents each.

So far as we could learn, none of the natives had ever penetrated the forest east of San Antonio to any greater distance than was necessary to obtain material for the construction of their rude dwellings. They could give us no idea of the character of the country a mile inland, nor could they tell even approximately the distance to the Jamary River, which was believed to be not far to the eastward.

We had heard tales of savages inhabiting this region but simply disbelieved them. Singly, or in parties of two or three, many of us had daily traveled miles over the cut lines to the south of San Antonio without being molested or seeing anything to indicate the existence of *Indios barbaros*, as the savages were called to distinguish them from the domesticated Indians, though Englishmen were said to have been attacked by them at San Antonio several years before. The immunity we had enjoyed from such visitations caused us to discredit all such statements entirely and to classify them with numerous pieces of published misinformation regarding the country.

This sense of security was slightly disturbed about this time. One Sunday, two men, out of curiosity, had penetrated the forest to a distance of about two miles back of San Antonio, when, lying at their feet, they found a vine, such as the natives use for tying together the framework of their dwellings. The piece of vine was at least fifty feet in length, freshly cut, carefully coiled, wrapped, and tied. Convinced that no inhabitants of San Antonio had ever been there before them, the two men took to their heels and never returned to the same locality again. Even this incident, when related at headquarters, created no uneasiness. A few days later, on April 17, some excitement was caused by workmen on the line circulating the report that they had seen savages, who bore no resemblance whatever to the semicivilized Indians employed on the railway.

On April 26, the Bolivians, Ignacio Arauz, and Santos Mercado, signed a contract to clear the entire line from the end of the tenth mile to the southern terminus of the railway for six hundred dollars per mile and to supply 2,500 crossties per mile of road at twenty cents each, delivered on the line. The clearing, as stipulated in the Huff contract, was to have a uniform width of one hundred feet.

On May 8, another contract was executed by Mr. Collins and Darwin H. Daniels, by which the latter agreed to furnish at twenty cents each, one hundred thousand hardwood crossties within one year and four months from that date and to supply all firewood required by the engines at $2.50 per cord, the amount not to exceed ten cords per day. It is believed that several more subcontracts for clearing and grading were made by Mr. Collins about this time, but of these no record is known to exist. The duties of the writer then confined him most of the time to headquarters, but occasionally, when some special service required it, he was engaged on the line and had opportunities to observe the progress of the work and life in camp.

On May 17, he was ordered by the chief engineer to go on foot over the finally located line, convey a message to Mr. Byers, and investigate an alleged error in levels. Byers was then camped some distance east of the Madeira, about eleven miles south of San Antonio and, with his corps, was engaged in establishing the final location over Mr. Runk's preliminary line south of Macacos.

The writer set out at 3:00 p.m., crossed the small stream a short distance south of San Antonio on a recently constructed and substantial rosewood trestle, and for one mile, followed a track then being laid on what was generally known as the "Shoo Fly." This term was then used by American railway engineers to indicate a piece of temporary track, constructed around unusually heavy permanent work, which would require much time to complete or to avoid some serious obstacle, the removal of which would otherwise occasion serious delay in transporting materials and provisions to laborers beyond. The expression was originally suggested by Ben Butler in Congress, when, in response to an interruption by "Sunset" Cox, he dismissed his diminutive antagonist with a wave of his hand and the quotation "Shoo fly! Don't bother me."

Strange to say, during subsequent litigation in the English courts, neither the distinguished judges nor the American witnesses could satisfactorily explain the term and the official records of the case today contain several references to the "Shoe Fly." The particular "Shoo Fly," here referred to, extended for a distance of two miles from San Antonio and was built in order that some heavy grading on that part of the permanent line might not unnecessarily delay transportation of provisions and materials to laborers employed, or about to be employed, on work farther south.

At Camp Lemon, three miles from San Antonio, the writer found men engaged in making a deep cut, and this was the extreme limit of grading then in progress. At the end of another mile, he stopped for the night with Mr. F. H. Clement, the resident engineer on construction, and found with him, as assistants, A. P. Scull, W. G. Coughlin, and T. C. Maher. We all obtained our meals at a woodchopper's camp in a hollow nearby. After breakfast the next morning, the lonesome tramp was continued for three-quarters of a mile farther, when, hearing that Messrs. Stewart and Ward were at the camp of Mr. S. B. Coughlin, eight hundred feet to the right of the line, we thought it advisable to see these gentlemen, because they were the persons who had reported the error in levels that rendered it necessary to send a messenger to Mr. Byers. Mr. Coughlin had a small contract for clearing and grading. For probably half a mile beyond his camp, the line was cleared to the full width of one hundred feet, but grading had not been commenced. Beyond was only the roughly cut line made by the engineers, frequently branching into side lines cut and surveyed to obtain the topographical features of country in the vicinity. Had it not been for familiarity with these lines, obtained from maps in the office, there would have been a constant danger of losing one's way. Every step had to be taken with care in order to avoid stumbling over stumps of young trees and bushes. The air was sultry, and not a breath of wind stirred the trees. Sometimes, no drinking water could be found for long distances. Consequently, progress was very slow and the journey tiresome. At one point, woodchoppers called attention to the sap of a tree, which they claimed was a remarkable curative for wounds and bruises. This tree proved to be of the species from which balsam copaiba is obtained. Further on, we encountered another kind of tree from which the Indians cut bark for use as cigarette paper. The forest troops of monkeys, swinging themselves by their long tails from tree to tree, so convulsed the writer with laughter at their antics that he could not even shoot straight enough to hit one. Occasionally, we saw wild turkeys. Macaws, in pairs, with plumage beautiful beyond description, flew screaming and screeching far above the highest treetops. Parrots and toucans came nearer but generally kept out of gunshot range, while every now and then, frightened peccaries would dart away with the speed of a jackrabbit from the cut line, where they had been indulging in a sun bath. Once, an opportunity occurred to shoot a deer at such close range that it seemed impossible to miss, but a small sapling, not more than an inch in diameter, though hard as iron, deflected the bullet from a .55-caliber Sharp's carbine to such an extent that the animal was not even wounded. In another place, while sitting on a tree trunk to rest, the writer was startled to find some half dozen Indians from Byers's camp almost on top of him before he was aware of their presence. In the almost impenetrable forests, where an American could be heard for half a mile tearing his way, these natives would glide along as noiselessly as shadows.

Soon the lid of a wooden soap box nailed to a tree and bearing the inscription "2 miles to Poverty Flat" was discovered, and, at 1:00 p.m., the writer found himself and Byers's cook the only occupants of that somewhat famous camp. It had required seven hours' actual time to traverse the distance of eleven miles. Around the supper table that night, many an incident of life in camp and on the line was rehearsed for the benefit of the unexpected guest.

Frequently, conversation reverted to what Byers's men spoke of as "a battle with the ants" on the previous night. As the ants had routed the whole corps in less than five minutes, the word *battle* seemed misapplied. Few of our men were well informed in regard to the wild animals encountered on the line, and the axmen generally gave them names based upon real or fancied resemblances to animals with which they were familiar at home. The peccaries were generally called "groundhogs" and experience in meeting them frequently on the line led all Byers's corps, very naturally, to consider them as harmless as rabbits. They were not aware of the fact that when encountered in large numbers, these creatures are dangerous and will tear a man to pieces unless he gets out of their reach by climbing a tree.

One of Byers's assistant engineers had caused the cook to complain by repeatedly coming back at night with his men too late for the regular evening meal. Byers had cautioned the offender in future not to permit his zeal for the work to prevent his return in time to join the others at supper. The very next evening, the same man was not only late, but much later than usual. On being called to account, he explained that he had been attacked by a drove of "groundhogs" and compelled to take to a tree and there remain until the siege was abandoned. This caused a burst of hilarious merriment, which the delinquent entirely failed to appreciate, and ever after, no matter when or where, if a person was asked whether he knew Mr. Blank, the answer would come after a minute's reflection, "Oh, yes! He is the man who was treed by groundhogs at Poverty Flat."

By 8:00 p.m. on the nineteenth, the writer was back in San Antonio, and from that time until the evening of the twenty-fifth, he was engaged with Mr. J. S. Ward in running check levels over the first six miles of the located line. While thus engaged, we had both suffered from severe attacks of chills and fever and, returning to San Antonio on the evening of the twenty-fifth, soon found that our experience had not been exceptional. Nearly every man there was or had been, within a very few days, sick or indisposed. It is a common thing to hear persons quote the relative death rate of two places as conclusive evidence of facility or difficulty in executing engineering work.

Nothing could be more fallacious than inferences drawn from such data. From an engineering point of view, the important question is not how many die, but how many are incapacitated for the performance of duty in a given time. The death rate may be extremely low and at the same time the number of persons unfit to work almost incredible. This remark applies with peculiar force to work conducted in places where malignant malarial fevers are prevalent. It requires a long time and frequent attacks of this disease to kill a previously healthy man, where proper diet and good medical attention are obtainable, but the very first congestive chill followed by fever often utterly prostrates the victim and leaves him more or less an invalid for weeks afterward. The large doses of quinine, required to cure the fever, incidentally increase the weakness and general debility. A period of lassitude follows, accompanied nightly by profuse perspiration, and all the time, neither the patient himself nor his companions regard the case as serious, though weeks must elapse, even under the most favorable circumstances, before he recovers his normal efficiency.

On May 28, Resident Engineer Nichols wrote to Colonel Church, "The contractors have about 300 men actually at work on excavation, embankment, clearing, etc."

From May 26 to 29 inclusive, there is a significant failure upon the part of the writer to make any entry in his diary. On the thirtieth, there is the following statement:

Am still sick and unable to do anything. Saw them carry a man to his last resting place yesterday. Noticed several others being carried, but could not tell whether dead or not. Just to give an idea of the sickness prevailing, shall mention such cases as have come to my notice within the past few days. I know, however, little or nothing about the laborers, being closely confined to headquarters myself.

Thomas Collins, chief contractor, recovering from fever; O. F. Nichols, resident engineer, sick a few days ago; don't know the cause; C. M. Bird, chief engineer, not yet recovered from effects of a trip up the Jaci Paraná; C. W. Buchholz, principal assistant engineer, has had slight attack of fever; C. S. d'Invilliers, chief of corps, recovering from an attack of dysentery; Charles F. King, leading sub-contractor, just over bad attack of fever; James F. Brown, chief clerk, recovering from something, don't know what; Hugh Kehoe, sub-contractor, had fever; Thomas J. Fetterman, druggist, just recovering from fever; Captain W. L. Symmes, in charge of river transportation, just recovered from fever; Captain Elder, clerk, very weak from fever; Fred B. Ealer, clerk, bilious fever; William S. Eustis, in charge of instrument repairs, fever; R. E. Johnson, Levelman, fever; J. H. Hiestand, chairman, recovering from his third attack of fever; Frank Snyder, chairman, has had dysentery a long time; McCracken had fever several times; George Yohe, headquarters' enterer, has fever; Mike, head-quarters' cook, fever; W. A. Rawle, clerk, dysentery; De Coursey, clerk, terribly emaciated don't know cause; McClellan Hirsh, mechanical engineer and machinist, just over light attack of fever; Robert B. Evans, chairman, dysentery.

All of Captain Stiles's party at the Caldeirão do Inferno are said to look like skeletons. H. B. Gill and J. W. Clark are said to have no shadows. Five men on Mr. Byers's party sick, some only slightly. Several of Mr. Runk's men are also reported sick, among them McIlvaine and Hoff—four in hammocks and two unable to be about. This statement indicates but a small fraction of the sickness we have.

On May 31, Paulino von Hoonholtz signed a contract to furnish on the line of the railway five hundred good laboring men from the province of Ceara, Brazil, at $1.50 per diem each, with provisions and medical attendance. This contract specified that the hours of labor should be from 6:00 to 11:30 a.m. and from 1:30 to 5:30 p.m. The same day, John S. Cox, transitman, was brought to headquarters in a hammock tied to a long pole and carried by two Indians. He had been with Mr. Runk's party. He was reduced to a mere shadow of his former self and was so extremely weak that he could hardly lift his head to drink a glass of water.

On June 1, a number of invalids were sent home. Provisions were again reported to be nearly exhausted, and Mr. Ward, who returned from Mr. Clements's camp on the second, reported seven out of fourteen woodchoppers there as unfit for duty. On the third, we were informed that the supply of flour was entirely exhausted, and on the same day, it was reported that Messrs. Runk and Stiles had been displaced.

Mr. O. F. Nichols, who had recovered from a slight attack of fever, was reported on the fifth to be worse than before. T.C. Maher, from Clements's camp, and W. H. Delleker, of Byers's party, returned to headquarters sick with fever. Other cases of sickness were constantly reported. Hugh Kehoe died on the sixth. This made five deaths that had occurred among our men, every one of them originally passengers on the *City of Richmond*, who had only been in South America three months.

On the seventh, Kerr, a woodchopper, came to San Antonio from a camp four miles up the line, quite unwell himself and reporting many of his men sick with fever. The condition of Hiestand and Eustis at headquarters was alarming, and Mrs. Packer, who kept a boardinghouse for laborers, reported seven sick men at her place. It was said that provisions could only last five days and that in ten days, unless the long-expected fresh supply arrived, there would not be enough eatables in San Antonio, either in possession of the contractors or at the native stores, to keep a mouse alive. It is not surprising that, under such circumstances, there should be many expressions of dissatisfaction and many complaints by our unpaid and half-fed men.

The three ladies at San Antonio—Mrs. Collins, Mrs. Nichols, and Mrs. King—set an example of uncomplaining fortitude, which did much to repress anarchic threats and wild talk among some of the laborers, who imagined that those at headquarters were faring better than themselves. On the morning of the eighth, when the writer entered the office to resume work, the chief engineer suggested his devoting the day to an effort to catch fish or kill some game. Accordingly, with fishing line and rifle, he wandered along the riverbank for several miles below San Antonio but would have returned at night entirely empty-handed had he not encountered some Indians, who sold him a number of fish killed with bows and arrows. That day, Mr. Brisbin from Mr. Byers's party, arrived at headquarters, unwell himself, and reported one engineer and two axmen were very sick at a new camp Byers had established some distance south of Poverty Flat. About the same time, Captain Stiles, with Lorenz, Gill, McCutcheon, O'Connor, and James Dougherty of his corps, returned from the Caldeirão do Inferno, all of them, except Stiles, sick and Lorenz and Gill in a dangerous condition. Captain Stiles said that at one time, out of his entire corps, he could not muster more than three men fit for service and that when the others were delirious with fever or crazed in consequence of the privations they had undergone, he could neither furnish them with proper food nor supply them with even such indispensable medicine as quinine.

Mr. John P. O'Connor, now private secretary to Archbishop Ireland, was then attached to Corps No. 4; he kept a most excellent diary and a digression is made in the next chapter in order to lay before the reader such parts of it as refer to the two months spent on the Upper Madeira with Captain Stiles. Another diary kept by Mr. Cecil A. Preston, now superintendent of the Middle Division of the Pennsylvania Railroad, covers the same period with much less detail, but supplements well Mr. O'Connor's story in several points where additional explanation is desirable. The exact phraseology of

Mr. O'Connor's diary has not always been preserved, but in no case has the sense been changed or a fact stated that is not to be found in the original. With this qualification, the authorship of the following chapter may be attributed to Mr. O'Connor himself.

XIX. Two Months with Captain Stiles

Yes, those days are now forgotten;
God be thanked! Men can forget.
—Adelaide Procter

On the morning of April 2, 1878, we began our preparations to move camp from San Antonio to a point about twenty-two miles up the Rio Madeira. With the help of a dozen Indians, all provisions and heavy baggage were moved to a landing place at a safe distance above the rapids of San Antonio. Gray and I were placed on guard the first night, and on the following morning (April 3), the tents were taken down and moved to the landing, where everything was stowed away in two canoes. At 8.30 a.m., everything being shipshape, we were ready to start, but something occurred to delay the Indians, and for a time, we rested as well as our cramped positions in the boats would permit.

At 10:00 a.m., we started, the eight Indians with their paddles keeping time with the puffing of the little steam launch, which went ahead and helped to pull us over the many rapids we encountered on our trip. Lunch consisted of some bread and hardtack, moistened with the murky water of the Madeira, after which we enjoyed a smoke from our long-stemmed pipes. At 4:00 p.m., after hard paddling by the Indians, we arrived at the Falls of Theotonio, where the boats had to be unloaded and both boats and cargo conveyed overland to a point above the falls, where we found a house constructed for our accommodation. It consisted of nothing more than a roof of palm leaves supported by posts about six feet long. Ends and sides were alike open to the weather and the passing stranger. Under this shelter, we all—sixteen engineers and sixteen Indians—swung our hammocks and rested our boat-cramped limbs for the night.

The next day was spent in moving stores, baggage, and boats by a rough pathway over the rising ground, formed by a ridge of rocks that extends directly across the river and causes all the tremendous disturbance heard by the traveler for miles above and below. The falls proper are about twenty feet high, and at this time, the rocks beneath them are invisible, the river being still near its maximum stage. About fifty feet from the east bank is a small island, making a separate—and apparently, the highest—fall on that side of the river. When the work of the day was over, many of us sat upon the rocks and cooled our heated brows under the falling waters. The arduous labors of the day gave the men good appetites for the rather scanty evening meal, after which we retired to our hammocks and puffed curling wreathes of Lone Jack tobacco smoke against the palm-thatched roof above us.

The morning of April 5 found many of the men at an early hour roving about the woods, some with rifles hunting game, others searching for fruit, and some enjoying themselves at the falls, watching the waters tumble down from the heights above to mingle with the circling eddies below, before starting on their long journey to join the mighty Amazon and be finally lost on the endless shores of the Atlantic. A few of the men, sick or tired, remained under the palm-thatched shed. One of the Indians and I climbed among the rocks overlooking the falls and were so fortunate as to find some wild pineapples, which destroyed our appetite for dinner. In the afternoon, Chief Engineer Bird, Principal Assistant Engineer Buchholz, and Resident Engineer Nichols, with a few others, arrived at the steam launch. We pulled the launch on shore, moved its freight above the falls, and made room for the new arrivals in a corner of our abode, where they, too, swung their hammocks. Then our day's work done, we all sought repose.

Early the next morning, April 6, we could be seen pulling hard on a rope fastened to the steam launch, and long before the sun had reached his highest altitude, we had the launch floating on the quiet water above the falls. Then came the work of reloading. Soon, everything was made ready, and an hour or so before noon, we were on our way. Twice, we encountered rapids and at one time had to paddle to shore, where the *Americanos* disembarked in order to lighten the boat, thus enabling the Indians to paddle through the swift waters about two hundred yards above where we landed. The Indians then paddled close to shore, allowing us to resume our seats and escape from the thick ant-covered underbrush, which lined the banks of the river. About dusk, we crossed the Morrinhos rapids, and just above them, opposite an island, we encamped under a tent, which, on account of the approaching darkness, had to be hastily erected.

An effort was then made to find solace for the privations of the day in coffee, hardtack, and tobacco, but the millions of insects about us did little to encourage hopes of obtaining restful sleep. The mosquitoes kept up a continual buzzing and biting. Smoking did not seem to have the least effect upon them, and our mosquito bars—traps would be a more appropriate name—did not help us in the slightest degree, but at a late hour, we became too much exhausted to remain longer awake even under such tortures.

In spite of it all, the morning of April 7 dawned upon us and rather late too; for, through the dense foliage above and around us, everything appeared gloomy and in complete accord with the feelings of those who, for the first time, began to realize that fighting insects by day afforded more physical relaxation than being their half-unconscious and helpless victims at night. Though the day was Sunday, we went to work with a will. Not until sunset did we find opportunity to rest. Where our tents were erected, not thirty yards from the riverbank, the density of the forest was such that no one, not even the native Indian, would move ten feet without using a knife to cut his way through the web of vines entangled with growing limbs of every kind—some thorny, others sticky and greasy to the touch, but all alive with red and black ants of all sizes. Some of these insects had an appearance of two pinheads fastened together by threads, while others measured fully an inch and a half and some even two inches in length. There were clods of flies of every shape and color and millions of winged and creeping insects, each and every one an expert in biting and stinging, ever ready and willing to bite and sting again.

About sundown, our labors were rewarded by the sight of all our tents in position and a considerable part of the campground cleared of vegetation. It was the hardest day's work ever accomplished by all of us, as all were forced to acknowledge.

During the progress of the work, we were scattered in groups, everyone as busy as the ants that bit us. Some were building fires to burn the damp and green brush, smoke ants out of their hills, or burn the nests of other bloodsucking and tormenting insects. Others were engaged in cutting poles, while more were employed clearing away large trees felled by our axmen.

A number found occupation in unfolding and erecting the tents, fitting in the ridge and end poles, and tying the sides and ends with short ropes to stakes driven in the ground. We made the stakes and poles ourselves. All things for camp use, such as shelves, seats, and benches, were made from nature's storehouse, the wild and untrodden forest around us.

The Indians did their full share of the work and proved apt scholars, continuing their labor with unabated zeal throughout the day. We all agreed that the dark-skinned sons of the South American backwoods had acted well their parts and much more humanely toward the pale-faced pioneers than would their copper-colored brethren of North America have done under similar circumstances. A plunge and swim in the river refreshed many of us, and, after some warm biscuits and coffee, followed by a few cigarettes, we retired to our hammocks, seeking the repose as much needed at the close of our first day at camp-making in the wilderness. Once during the night, ten of my mates and I were awakened rather suddenly to the pole, which held one end of our hammocks, giving way and throwing us to the ground, where we received many an ant bite before lights were provided and our situation realized.

At the late hour of eight o'clock in the morning, we left our hammocks, even then reluctantly and longing for another turn over and another quiet doze. Our work of the previous day had to be completed, and we went at it vigorously, cleared away more thoroughly the small roots and leaves, raked over the ground within and without our tents, and swept every place until the entire camp looked as neat and orderly as though the work had been executed under the eye of some industrious housewife. The rake and brooms were made by those who used them. Clothes lines were made from vines that hung from almost every limb, and the hooks, to which they were fastened, were found growing on the trees. During the afternoon, the *Americanos* rested. The Indians did a little work to pass away the time until night brought some degree of rest and quiet to all.

At sunrise on April 9, we were all seated at the breakfast table, and after the meal had been disposed, we began our line a few yards back of camp, using the surface of the Madeira as an assumed datum plane for our levels until a connection with Byers's line would enable us to adopt the San Antonio datum. After running directly inland to a given elevation, in order to reach the place where the finally adopted railway location was expected to pass, we were to change our course to N.24°W., which it had been estimated, would bring us to a connection with the party running toward us from below.

Not two hundred feet from the starting point, we had to

cross a creek twenty feet wide and ten feet deep. It required just four and a half minutes for three Indians to bridge that stream. They did their work well, and the bridge will last as long as the timber in it. They cut down four small trees, eight to ten inches in diameter, laid them from bank to bank, bound them well together with vines, and secured the whole to four long poles driven a couple of feet into the ground. A vine was then stretched between these poles on each side of the bridge to serve as a handrail. In the same way, they bridged a number of other streams on our route.

One day, while we were enjoying our luncheon in the woods, a score of monkeys on the treetops, one hundred and fifty feet above us, were looking on and, I suppose, chattering among themselves on the delights of munching hardtack and satiating one's appetite with dainty morsels of "salt horse," in which the *Americanos* were busily engaged. A rifle shot soon put an end to their speculations. One of them dropped, and the others took to their heels and tails, for they used their tails rather more than their other limbs during their travels.

The afternoon of the same day, Chief Engineer Bird put in an appearance. After raising a little breeze for the benefit of our captain and pertly inquiring whether he was "going to Rio," he ordered all hands to return to camp, which we did. The line, so far, had reached a distance of about three and an eighth miles inland and passed over a country, which, except for a few creeks we had to cross, was quite regular and without abrupt or precipitous slopes. The growth of tiny vines and underbrush was the thickest and most entangled I had ever seen. No vegetation about which I have ever read or even dreamed could compare with these in density. The axmen and vine cutters ahead had to chop and cut at every step, so that the remainder of the party could see their way through and follow. Only on very rare occasions could we see twenty yards ahead of us, and a man going that distance from the party would have to use his voice to determine the position of his companions and the direction in which to return.

Though extremely large trees were rarely found on our route, it was seldom possible to see the topmost branches of any. The constant struggle for sunlight between the trees and the parasitic vines, which climbed over them, made the foliage near the treetops almost or altogether impenetrable to the eye.

The largest trees we saw were not more than twelve or fifteen feet in diameter and two hundred or more in height. The vines twined around every tree, large and small, in most cases binding several together. Every tree had a thousand roots below and a thousand branches one hundred feet above, with many times that number of seemingly endless vines climbing up or hanging down and connecting the leaf-growing branches above with the life-giving roots below. I may truthfully add, in regard to insects, that every branch and uncovered root is thronged thoroughly by poisonous ants and other insects of various sizes and colors. Every leaf of the foliage above us had its quota of little insects. Such was the countless number of living creatures in this yet unknown and almost impenetrable forest.

On Wednesday, April 17, orders were issued to move camp to station 115, 2.2 miles from our old camp, on the line bearing S. 45°E. By 8:00 a.m., the men could be seen wending their way along the line, each with his hammock and a few changes of clothes slung over his shoulders from the end of a .55-caliber Sharp's cavalry carbine. The line was easily followed, for nowhere else was there a clear pathway. We reached our journey's end in due time, and all immediately went to work with ax and knife to prepare the new camp.

The flies soon discovered that we were intruding and made such a raid upon us that for a time we despaired of ever finishing our work. They crowded so thickly over us that it was almost impossible to see the color of our clothing, and as we warmed to the task, from our heads down to our boot heels, we were enveloped in one grand living coat of gold-colored flies. Truly, we were the possessors of millions of flies. One blessing, for which we were duly grateful, was that they did not bite. We called them "sweat flies" and made the best of a bad situation by not continually trying to brush them off.[19]

The Indians with tents, baggage, and cooking utensils arrived on time, but through some mismanagement, a number of necessary tools could not be found, so we were unable to move and complete the establishment of our new camp in one day, though we were not much inclined to do so. For shelter during the night, we rigged two flies belonging to the large fourteen-by-fourteen tents, using a stout vine as a ridge pole, and four trees, which a Good Providence left conveniently

---

[19] I never saw so many flies in my life ... It was a pure hell, but we did what work we could ... Some of the men lay awake all night on the lookout for tigers and Caripunas. We could hear tigers around us ... The flies are terrible and will be brushed off ... I feel more like returning to Philadelphia since I have been here, than any time in the last three months," from a diary of Cecil A. Preston, CE.

in our way, served as corner posts. Two strong poles, thirty feet in length, were stretched lengthwise on each side at the proper height and fastened to the trees. To these, the lower edges of our flies were tied and our hammocks attached, thus completing the temporary abode for the reception of its weary builders. A few pieces of hardtack still remaining and some hot coffee cheered and refreshed the weary and downhearted. After making the most of this scanty fare, we retired to rest or to swing in our hammocks. Jokes and stories began to circulate, encouraged by the propinquity of too many persons crowded together under one canvas. Many a curse and many a laugh were mingled with comments upon our experience as compared with the romantic tales of others relating to life on the Amazon and Madeira or in the trackless primeval forests of tropical Brazil. Noises of all kinds kept us awake far into the night, notwithstanding our earnest efforts to find refuge in sleep from the unpleasant thoughts constantly suggested by our situation and surroundings. Hundreds of birds, with caws, screeches, and whistles of every key, kept up an accompaniment to the chattering of monkeys. Tigers seemed to be prowling and growling among the leaves at the foot of every tree. They could be heard tearing their way through canebrakes or splashing in the waters of a stream nearby, while other animals, of unknown species, added their squeaks and grunts to swell the discordant chorus around our camp. Thus it continued until long after midnight, when, the last joke cracked, the last story told, we finally became unconscious to the infernal music around us and sank into a troubled sleep.

The next morning, we did not linger long in our hammocks, but after a refreshing bath in the crystal waters of the before-mentioned creek, followed by a single cup of coffee—no hardtack this time—we were ready to erect our tents and complete the work upon our new camp. We had a repetition of the previous day's experience with flies, which continued to settle over us in undiminished numbers, but early in the afternoon, we finished the work about camp and were able to obtain a little rest and a smoke, while swinging in our hammocks. The next day, we continued cutting our way through the forest as usual.

On April 25, while eating luncheon on the line, we were once more honored by a visit from a dozen or more monkeys, who jumped and swung on their tails from branch to branch of the trees 150 feet above our heads. Three rifle shots dropped two of them, one a huge, black ring-tailed rascal, the other smaller and of gentler aspect. The remaining quickly utilized feet and tails to skip for the unknown.

On the afternoon of April 29, I was visited by the first attack of chills and fever while at work near station 47 of our experimental line, running N.24°W where we were just beginning an offset line for topographical purposes. From this attack, I did not recover until Saturday, May 4, when, feeling better, I resumed work, but I had another attack of chills and fever that night.

The ground passed over was very irregular. We had to cross several streams and gullies, some of them twenty feet deep and not quite that distance across. Large patches of wild pineapples, with their thorny leaves, and extensive canebrakes, interlaced with prickly ones, made even less attractive our usual unending contest with the dense vegetation and omnipresent ants and flies.

Returning to camp early on the afternoon of May 5, we were greeted with the—to us—joyful news that the corps had been ordered to discontinue work and move at once to the Cachoeira de Caldeirão do Inferno (Falls of the Boiler of Hell), some forty-five miles farther up the river. By the morning of the tenth, everything belonging to us had been conveyed to the falls and, with ourselves, had been packed in the canoes that were waiting.

The voyage up the river was quite interesting and the weather generally pleasant; only two light showers occurred during the trip. In addition to the corps of engineers and eleven Indians, we had on board my original fellow passenger on the *Richmond*, the writer, a representative of the *New York World*, and Mr. Mariano, the captain of our Indian crew.

Mariano, a lighthearted young fellow, about twenty-three years of age and a native of the valley of the Orinoco, was a very entertaining traveling companion. On the way, he told us the story of his trip by canoe from the Orinoco to San Antonio, and as he described the deep canyons through which rushed with great noise the swelling waters of that river, his face, no less than his eloquent language, gave evidence of the ardent love and patriotic devotion that still infused every thought and utterance concerning the land which gave him birth.

In a small canoe, accompanied only by one Indian boatman, he had paddled against the stiff current of the Orinoco, camping on shore and on islands at night. On the tenth day out, he entered a submerged country filled with many islets and rugged rocks. Continuing through this lake, he was finally caught by an eddy. It carried him into an outlet, which gradually widened into a well-defined stream flowing in a contrary direction to that of the Orinoco. Floating down it, they found it steadily increasing in size, passing through many dangerous rapids, where only the skill and dexterity of the Indian boatman saved their frail craft from destruction, until finally, they found themselves on the Rio Negro, which they followed to Manáos. There, Mariano made a present of the canoe to his faithful assistant and, bidding him farewell, took a river steamer down the Rio Negro to the Amazon, down the Amazon to the Madeira, and up at last to San Antonio.

Once during the day, we had to get out of the boat to enable the Indians to paddle through some small rapids and at night came to a stop at a place we named Dead Dog Station, where among some old, rickety untenanted houses we found, or rather smelled, some dead member of the canine species—hence, the name.

Continuing our journey the next day, we had to get out of the canoes again, this time on a log several yards from shore, and some of the men fell into the river. At 3:00 p.m., we reached La Concepcion, the finest plantation above Pará. The extensive clearing was carefully made and contained a good collection of clean, new-looking houses. Dwelling in these were several hundred Bolivian Indians with their families, officered Bolivians of Spanish descent, all under the control and in the employ of Don Pastor Oyola, who occupied the most substantial house and reigned with absolute but steady sway over, not only this plantation, but many smaller ones in the vicinity and many square miles of the richest rubber-producing forests in Brazil. Making our bow to the Don, we were received with the utmost cordiality and most hospitably entertained at a royal supper, after which cigarettes of the finest Bolivian tobacco, unequalled in flavor by any other, were puffed and wine flowed freely. The men thoroughly enjoyed themselves.

One object that attracted our attention was a large wooden mission cross standing near the riverbank in front of the plantation and facing inland. I now forget the date upon it and the name of the particular order in which it was erected. They belonged, however, to those good missioners, who, in every clime, from the scorching suns of the torrid zone to the mountains of perpetual ice and snow at the poles, with the cross as their emblem and sword, have ever been the sign of civilization and progress. The cross was about eighteen feet high, neatly painted black, and lettered in white; from the arms hung two pieces of woven sash about eight feet long and nine inches wide. At sundown, the entire population assembled in front of the sacred emblem, and after they reverently repeated their short prayers and attended to little devotions, a litany of the BVM ("Blessed Virgin Mary") was sung by the male and also by the female part of the congregation. Some gave short prayers and devotions, enlivened at times with music. Then came the singing of the "Alma Mater," which all joined in. When devotions were over, the choir, consisting of performers on two violins, two large drums made from half barrels, a few flutes, some steel triangles, and one United States kettledrum, assembled in front of the house where we were quartered and gave us a serenade. From another such serenade in future may good fortune deliver us!

Dancing and singing were next in order, and after these,

a sham bullfight took place, the bull being an Indian covered with the head and hide of an ox. This closed the program for our entertainment, and at 11:00 p.m., we all sought peaceful slumber.

The next day, though it was Sunday, we were moving before 6:00 a.m., still paddling against a strong current, which compelled us to keep close to shore. About 10:00 a.m. we landed on the riverbank, where the Indians breakfasted and the *Americanos* had some coffee, biscuits, and roasted yuccas. The last grew near to our stopping place. At 1:00 p.m., we passed a place on the riverbank, where in April 1875, five out of seven Bolivian traders were killed by the arrows of ten Indians on the bank, who pretended that they had come to trade with the victims. At 2:30 p.m., we crossed the east bank of the river and landed soon after at a plantation belonging to our host of the previous night. Warned by dark and threatening clouds that a wet night was in store for those without shelter, we waited to partake of a hastily prepared luncheon before hanging our hammocks under a convenient roof made available at our disposal.

During the afternoon and evening, an engineer and I shot some gaudily colored macaws and parrots. I also captured a few new specimens of butterflies that were added to the already large collection made by the *World* correspondent.

Early the next morning, we were again on our way. Life was on every side, above and even beneath us—dolphins and other fish in the river, macaws, parrots, and parakeets in the air and monkeys on the trees.

Many a rifle ball went astray on account of the motion of the boat. Nevertheless, a few monkeys and parrots could be seen on board and many dead or wounded animals were left behind. At 1:30 p.m., we arrived at the Cachoeira de Caldeiráo do Inferno and soon afterward entered our destined harbor, where we eagerly sprang ashore rejoicing to be free from our cramped positions in the canoes.

Nearby was a large plantation, belonging to Don Ignatio Arauz, who has many others on the Madeira and one on the Amazon.

Señor Arauz had preceded us on our trip up the river and was waiting to welcome us at his plantation. We were shown into a new house, which he had just built for the use of the railway engineers and contractors, and there we were quartered during the two weeks and one day following our arrival. While there, we witnessed the making of molasses and the husking of corn. On Wednesday morning, May 15, we slung our hammocks from our rifles and proceeded inland two and a seventh miles over an old trail, cut by locals for the English engineers a few years before. It terminated at another rude native house, where we established Camp No. 3. On the sixteenth, we returned downriver and began our line, following the old trail all of the way to our camp, and made the distance about seven hundred feet.

Early on the afternoon of the twentieth, after helping the Indians to bridge a small creek, I was again attacked by chills and fever; I was compelled to return to camp and there remain for some days, suffering little pain but steadily growing weaker and rapidly losing flesh. When, on the twenty-sixth, a fastening gave way and our tent fell in, some of the men were so weak that they did not have sufficient energy to move from the places where they fell. Probably the shock was beneficial.

Certainly, it put enough life into me to seek the bank of a little stream near camp and there indulge in a refreshing head and arm bath. Several others followed my example. Suddenly, some of those coming down the bank seemed to be seized with a panic and came tumbling and sliding down the steep slope on top of us. The cause of the commotion became apparent when a few seconds later, a snake about eight feet in length followed and disappeared near the water's edge.

Seeing a prospect of long woodland trips ahead that morning, I walked to the landing at the river to find that some extra baggage I had left there had been removed. On the return trip, good Stiles balanced me by giving me a small haversack to carry. Feeling better, I did not refuse, though my rifle alone was weight enough for a man in my extremely weak condition to carry.

By May 27, we were again ready, as some of the men expressed it, for "a change in locality, but free of neighbors," and all hands proceeded to the end of the cut line to clear the ground for a new camp. Soon, we had a considerable area denuded of vegetation and the brush piled, fired, and sending volumes of white smoke heavenward. About 2:00 p.m., we were back in our old camp and preparing to move the next day.

By the twenty-eighth nearly all of us were sick, and work on the line was consequently impossible. A few still remained unconquered, and they volunteered to go to the landing for flour and such necessaries as could there be procured. Thanks to one good sportsman, or some lucky shot, we had a turkey dinner. Oh, but it was delicious!

The next day, I was one of only four men who were able

to do some topographical work on the line between stations 5 and 25. About 11:00 a.m., while we were eating luncheon and taking a little rest, some fifteen or twenty young and old monkeys, black, brown, and gray, some with long tails and some with no tails at all, appeared in the trees above our heads and peeped down at us. We remained perfectly quiet, but one of the men made a roll of newspaper, which he partially hid among the leaves and branches some twenty yards away. Soon, curiosity led the monkeys to slide down the tree trunks to investigate the mystery. Four with long tails, closely followed by two others, sniffed at the paper. Suddenly, several long arms grabbed for it and an explosion of monkey chatter followed. A minute later, from the trees one hundred feet above our heads, they were closely inspecting the editorial columns of our only New York newspaper. Evidently, they were disappointed in the contents, for soon, the air was full of torn pieces of newspaper, and the monkeys sought to amuse themselves at our expense by tugging with all their might at the smaller branches of the trees in an attempt to shake something down upon us. As these particular trees bore no large nuts and had no dead branches, they exhausted their own strength without injury to us. About 2:00 p.m., we returned to camp, and an hour later, seven of the party were lying sick and weak in their hammocks.

The next day, camp was virtually turned into a hospital.

June 2
The past few days have been full of misery and suffering for the twelve men under Captain Stiles. Not less than ten have been in their hammocks all day, sick and crazy from chills

and fever. Fred (Lorenz) had a crazy fit last night. Many of us have become indifferent in regard to further employment on the work. At one time, two of the men were regular idiots, imagining all kinds of company crowding them out of their easy chairs. We all had nights of restlessness—plenty of nightmares and no repose.

Weakness and loss of vitality, due to insufficient food more than to any other cause, have reduced us to our present condition. There seems to be no remedy in sight. Stiles means well. Preston can do nothing and is as sick as any of the others. Eustis, who is an excellent shot, is but a shadow. Gill has some energy but not enough to carry him out of the clearing. One is stark crazy, and I am on the flat of a gone back. Our supplies, which we should have had with us when we left the river some two weeks ago, have not yet arrived.[20]

June 7

Another week of misery and mush. Yes, mush and molasses, molasses and mush, with fried mush and molasses for a change.

Stiles left for Don Pastor Oyola's plantation early in the week. We trust he left hunger behind him. On Wednesday, Buchholz tripped in with what was more welcome—provisions.

---

[20] "On the 5th of June, while visiting Corps No. 4 in camp 8 miles from Caldeiráo, seven of the eleven men composing the crew were down in their hammocks unable to work and had to be carried by Indians to the landing. I was obliged to send six of them to San Antonio to receive medical attendance. (During these months of June and July) I never visited a camp without finding three or four men on the sick list. Of the fifty men composing the different crews, I know of but two that escaped," from a letter of Charles W. Buchholz to Franklin B. Gowen.

Fisher came with him and will remain with our corps. A few hours later, we learned that we were to be *relieved*, not in the sense that insanity and idiocy relieve chills and fever. Six of our party left the Caldeiráo do Inferno this morning for headquarters, recuperation, and health. What a fever's ending after all! A few days ago, we were, as some of the men put it, a joint-faced, soul-sick, dumpish crowd, fit only to shake our bones in a hammock, but now we are, though still far from well, animated by hopes of improved conditions in the future.

As for experiences, we have had a variety, from cornmeal three times a day for weeks to monkey once in a lifetime. Monkey as an edible, by the way, is something not to be sneered at. We plead guilty! It happened in this way. Eustis, an expert with shotgun and rifle, had wandered from camp at a time when the stock of provisions was extremely low and we could supply no proper diet to the many lying sick in camp. We heard a shot. Hope gave us strength, and we prayed for more shots. We heard more. Time passed by, and the incident was nearly forgotten, when, late one afternoon, a suspicious odor was wafted near our tent. Even then, we ventured to build no castles in the air. Odors in the forest are not uncommon, but this one proved extremely gratifying to our olfactory nerves. Fred Lorenz thought he had a return of previous hallucinations when a few minutes later, the cook announced a feast in readiness for us. We toddled to the table. "What a savory smell! Where did he get it? Who shot it? What is it?" The chills were soon forgotten. The meat was tender, possibly a little strong, but prepared like young deer or tapir, it pleased and tickled our cornmeal-worn palates.

We enjoyed it. What it was only heaven and the cook could tell, and they would not. Threats proved useless. We ate long and heartily, called for more, and even after returning to our hammocks, we continued picking the meat and sucking the marrow from the bone. Later, one of our men, a little stronger than the others, ventured a short distance into the forest in the hope of being able to shoot *another* deer. There, he found two big black monkey hides with heads and tails attached, clearly explaining the origin of our feast—humanlike, but not of our species. We all came, we saw, and we were conquered. The discovery was too shocking for men in our debilitated condition to endure. We couldn't resist the rebellion within. The meat we so longed for and enjoyed was not ours to keep. Darwin may be right. Who knows? But, in future, let someone else than the writer claim life from a monkey or monkey's stew. The experience did us no harm, and we felt thankful that we had not knowingly been cannibalists.

About 6:00 p.m. on the seventh, after a pleasant trip down the river, we landed on a rock in the rapids above the Falls of Theotonio, where we camped for the night. For once, at least, we were free from bugs, flies, and swarms of gnats, but we were painfully conscious of the hardness of the rocks upon which we endeavored to pillow our heads and rest our emaciated limbs. It was unnecessary to call anyone on the morning of June 10. Our aching bones and stiff joints made us eager to depart, and at an early hour, we were on our way. About 8:00 a.m., we pulled ashore at San Antonio and once more breakfasted at headquarters. We were a sorry lot. The two months we had been away had wrought quite a change, not only in us, but also at San Antonio. Many of the men we left there on April 3 had departed for home, and there was evidence of no slight progress in the work. Such things, however, failed to interest us; we thought only of our chills and fevers, which continued with unabated frequency and violence.

XX. At Headquarters Again

Life is mostly froth and bubble,
Two things stand like stone—
Kindness in another's trouble,
Courage in your own.
—Gordon

During the period from June 9 to July 8, the continued inability of the contractors to pay their men and the inadequate supply of provisions had an extremely demoralizing effect upon the entire laboring force in the vicinity of San Antonio.

Under the contract signed by them before leaving Philadelphia, the men were charged with the cost of transportation to San Antonio until they had served six months and were only entitled to a free return passage at the end of two years of service. Thus, it happened that many of them, particularly those who had lost time on account of sickness, had no money between them. Many were even in debt to the contractor. Those to whom wages were due could purchase articles of clothing, tobacco, and a few other necessities at the company store conducted by P. and T. Collins, but no food could be obtained, by purchase or otherwise, suitable to the climate or calculated to tempt the appetite of the invalid. Orders were occasionally issued by the contractors, in limited amount, on the two small Brazilian stores, but these were of little use except to purchase intoxicating liquors, the supply of which was at times abundant. The immoderate use of these intoxicants invariably caused the prevalent fevers to recur with increased frequency and to assume an unusually malignant type. The men could obtain no cash for transmission to dependent families at home or to meet the expense of a return voyage, should they desire to abandon the enterprise. Most of the time, cornmeal, rice, and coffee were the only provisions obtainable. On June 12, the supply of wheat flour was exhausted and the force of laborers actually employed on the line was so small that the work of construction was practically suspended.

Early in July, the corps, under Mr. Byers, which had encamped about fifteen miles south of San Antonio, were forced to return to headquarters on account of inability to obtain provisions. Numerous hunting parties were organized for the purpose of obtaining occasional delicacies for the sick, but the supply of game thus obtained was entirely disproportionate to the labor involved in procuring it. We had no means of manufacturing ice, and without it, fresh meat could not be preserved longer than three or four hours, even if we were so fortunate as to obtain it. The extent to which we could rely upon game for our table supply was indicated by an incident that occurred on June 10, when Messrs. Bird, d'Invilliers, King, Patterson, Ward, and Stewart returned from a trip to the Falls of Theotonio, where they had gone on a hunting excursion extending from Saturday morning until Monday evening. The net result was about two dozen pigeons and exactly six subsequent cases of malarial fever, due to exposure while away.

There seemed to be, during the time specified, a decided improvement in the health of the American colony, and not a few of us supposed for a while that we were gradually becoming "acclimated." Many of us shared in the erroneous belief, prevalent among even generally well-informed persons at home, that after passing through a process of adaptation to new conditions and surroundings, we would eventually reach a state in which we would enjoy partial or entire immunity from malarial fevers. We learned later that the natives, under similar conditions, suffered from such diseases quite as much or even more than we did and that complete acclimatization was only obtained by those who acquired permanent subterranean abodes in the graveyard. The improvement noted was probably due not to any physical change in us or to any climatic variation so much as to fact that we were learning how to take care of ourselves, the kind and quantity of medicines to take, the particular edibles and beverages it was desirable to use or to abstain from, and certain precautions, relating to clothing, that must be observed by those who hope to maintain even moderately good health in malarial districts. It was simply impossible for men to thrive in such a climate upon the kind of food usually supplied to laborers at home. The sick required a most carefully selected diet, which we were unable to supply. At times, we considered ourselves fortunate if we could provide them with such delicacies as fat bacon, salt beef, and cornmeal.

On June 11, the last grain of quinine available at the drugstore was consumed, and our physicians began purchasing all that was obtainable from men who had been thoughtful enough to bring a private supply with them from Philadelphia. The amount thus procured was entirely unequal to the great demand for this indispensable febrifuge; in order that patients might not be discouraged by a knowledge of the truth, the doctors compounded and regularly administered fraudulent sugar-coated quinine pills that contained no quinine at all.[21]

At no time did we have a hospital that deserved the name. The little shanty used as such until June 8 would accommodate only five persons, and the most commodious space could scarcely contain their beds. The best establishment for the care of invalids that we ever possessed contained cots for only twenty-five.

The presiding officer of this beneficent institution, into which at one time seventy-six invalids were crowded, was a rough but good-hearted Irishman generally spoken of as "Old Mike," who, on account of a long and varied sick-care experience acquired in California during the famous days of the 1849 gold rush, was supposed to be preeminently qualified for performing the multifarious duties of his responsible position. Mike would cut a sorry figure in a group of trained hospital nurses today. He wasted no time upon his toilet and scorned the adventurous greatness that, without regard to

---

[21] "There were also times when the proper medicines were scarce; and never did have a sufficient amount of the indispensable *quinina*; after we have been for weeks without it altogether, and had to borrow from our friends, the *seringueiros*," from a letter of Charles W. Buchholz to Franklin B. Gowen.

personal merit or professional ability, usually attended the wearer of a uniform.

Attired in a blue flannel shirt, a pair of trousers belted in at the waist and much the worse for wear, high boots, and a dilapidated straw hat, he was at all times equally well prepared to dose the dying or dig a grave for the dead.

The discoloration of his gray beard made plainly evident the only vice to which he was addicted, while his powerful frame, combined with a stern and naturally reticent manner, impressed even a casual acquaintance with the idea that he was not a person with whom it would be safe to trifle. It is extremely doubtful whether Mike could distinguish Epsom salts from tartar emetic, and the attending physicians would just as soon have thought of asking him to calculate a table of logarithms as to use a hospital thermometer. Convinced that strict obedience to orders was the great secret of success in hospital practice, Mike never failed to administer medicine at the exact hour stated in the prescription. The number of invalids in camp and hospital was too large to permit the waste of valuable time in coaxing and cajoling whimsical patients.

On one occasion, a man, who was so weak that it seemed each breath might be his last, objected to taking his ration of fraudulent quinine pills. In an instant, a ponderous fist was raised above the patient's head, and with a "D——n you, take them," from Mike, the pellets of sugar-coated flour paste disappeared. Whether the particular individual survived the treatment or not, the writer fails to recollect, but not long after, he observed Mike, assisted by a common laborer, busily engaged in constructing a rude coffin for some unfortunate victim, digging a grave in the banana patch, and subsequently conducting the funeral services with a devoutness of manner that could not be excelled by any prelate in the church. "Old Mike" had his limitations, but no man of all those who went to Brazil in 1878 was more faithful in the performance of what he considered to be his duty.

As superintendent of the hospital, coffin-maker, funeral director, gravedigger, and chaplain, he was a prominent character, not only at headquarters, but later on his voyage to Pará and during his stay in that city.

It was impossible for him to supply the deficiency in hospital accommodations, provisions, and medicines, but he utilized what knowledge he possessed and every resource at his command in a prolonged and earnest effort to alleviate the sufferings of the sick and in a measure atone for the "lack of woman's nursing" and the "dearth of woman's tears" at many a deathbed in San Antonio. In subsequent years, the writer has seen many trained nurses and some trained nuisances, who, for a valuable consideration, endeavored to mitigate the horrors of the hospital by ministering to the wants of the invalid or by posing before him in becoming costume, but with all their skill and training, they have utterly failed to banish the unique figure of "Old Mike" from a place among his fondly cherished memories of the past or in the slightest degree to diminish the profound respect with which he has always regarded Mike's unselfish and unrewarded devotion to duty.

Neither the sterling virtues nor the strenuous character of its superintendent could, however, make the San Antonio hospital a popular resort for invalids, and the reader will not find it difficult to believe that few not delirious with fever or too weak to resist were ever carried there. By far, the greater number of invalids never sought medical advice so long as they could obtain quinine without a doctor's prescription. They simply remained in their hammocks at headquarters or in camp, dependent upon companions for what little attention they received. Thus, it came about that the official medical reports failed to indicate even approximately the amount of sickness in the immediate vicinity of San Antonio. Our physicians, so far as the writer can recollect, never went more than a mile or two from headquarters, while our camps extended as far south as the Caldeirão do Inferno, a distance of about seventy-six miles by the river.

Though the writer was closely confined to the offices at headquarters, where his opportunities for observation were very much restricted, his diary mentions from June 10 to July 7, no less than thirty-three new cases of sickness, three of which had a fatal termination. The list of those prostrated by fever includes the names of many of the most prominent men connected with the enterprise. Mr. S. B. Coughlin, formerly storekeeper but then a contractor, was carried to San Antonio in such a condition that for a time it was doubtful whether he would ever recover, and he finally escaped alive only by abandoning his contract and returning to Philadelphia with a number of others in an equally bad condition.

On June 28, Dr. Townsend announced that there were 186 men at San Antonio whom he regarded as disqualified by their physical condition for further service in that climate.[22]

---

[22] "On the 21st of June the doctors reported 260 men on the sick list, out of a total of 550 then working within four miles of San Antonio," from a letter of Charles W. Buchholz to Franklin B. Gowen.

Though the maximum daily temperature was no higher than is frequently observed in our Southern states during the summer months, the daily range of the thermometer was great and constantly increasing as the season advanced. The atmosphere was at all times so saturated with moisture that the usual fall in temperature at night was invariably accompanied by a very heavy precipitation of dew. The highest and lowest temperatures observed on a Fahrenheit thermometer in the shade, between March 23 and June 22, were from 58 degrees to 97 degrees, but the minimum daily temperature, which, of course, was always reached before daylight in the morning, was much lower during the month of June than at the time of our arrival in February.[23]

Early in June, the situation became so desperate at San Antonio that Mr. Charles L. Moore was dispatched to Pará to ascertain the cause of delay in the arrival of supplies and, if possible, to remedy the trouble, whatever it might be. Among the laborers, there was much wild talk. It was even said that an attack might be made upon the engineers at headquarters by Italian laborers and others, who suspected that we had more and better food than they did. A proposal to seize and hold the Amazon Navigation Company steamer, due June 15, in order to have a means of escape when provisions were completely exhausted, was openly discussed.

About June 11, Chief Engineer "Charlie" Bird began the construction of a light rowboat, which could be easily carried

---

[23] "At San Carlos, on the 17th of May the thermometer fell to 50 degrees at 6 A.M. and frequently went down to 60°. On July 9th the mercury at San Antonio fell to 60° at 6 A.M. and never rose much above that during the whole day," letter of Charles W. Buchholz to Franklin B. Gowen.

around obstructions by two men and be used in exploring small streams, which neither the large canoes in use on the river, nor our steam launch could enter. This unfortunate coincidence added fuel to the flames and gave rise to a report that the chief engineer was preparing to abandon the enterprise and make his way as inexpensively as possible to the seacoast. (Charlie Bird's boat became part of a later legend, known well among the natives of the lower Amazon River Complex.)

About June 13, Mr. John Runk, who had been engaged in running the preliminary line from a point two miles southeast of San Carlos toward Las Pedras on the Jaci Paraná, was relieved from duty, and Mr. Robert H. Bruce was ordered to assume charge of Corps No. 3. Two days later, Mr. Runk arrived at San Antonio and there joined Captain Stiles in waiting for transportation to Philadelphia. It was generally understood that the service of neither of these gentlemen had proved satisfactory to their immediate superiors, and it is hardly necessary to add that the harsh conditions under which they were expected to work proved equally unsatisfactory to them. On the thirtieth, the long-expected steamer *Arary* arrived from Pará. For the first time, she anchored out in the river instead of coming up to the bank, and this led to the suspicion that some benevolent person had posted her offshore, probably regarding the plans that had been discussed for capturing her. She brought no provisions worth mention, but her officers relieved our anxiety to a great extent by telling us that two schooners laden with supplies had arrived at Pará and were waiting to be towed to San Antonio. In consequence of this assurance that relief would reach us at an early date, our depressed spirits once more revived and attention was again concentrated upon plans for the earnest prosecution of the work.

Intimations received in private letters from home, together with our knowledge of the local situation, convinced most of us that the contractors were and must continue to be financially embarrassed until the payments for work done and material furnished were sufficient to compensate them for the enormous initial expenditures involved in the execution of their contract. Their estimated earnings as yet fell far short of the amount necessary to pay expenses. This state of affairs was not, however, unusual in the initiation of contract work and caused no serious apprehension relating to the ultimate financial stability of the contracting parties. It was expected that the payments, then due the contractors from the Madeira and Mamoré Railway Company, would be sufficient to reestablish their credit and enable them to pay their employees and meet some other pressing financial obligations. There was no reason to doubt that the money in the Bank of England would be promptly applied to the purposes for which it was raised, and it seemed to be only a question of time, when earnings from work, contemplated and in progress, would dispel the clouds that obscured the financial situation.

The badly crippled remnant of Captain Stiles's crew, consisting of six Americans and four Indians inclusive of Mr. Cecil A. Preston, was still encamped near the Caldeirão do Inferno awaiting provisions and instruments that would enable them to continue their survey northward to the banks of the Jaci Paraná, a small, but navigable, stream that empties into the Madeira about fifty-two miles above San Antonio.

To place this corps in a serviceable condition, Roman

McIlvaine, Arthur P. Herbert, George A. Schele, John Delario, Milton Brown, James T. Young (from Johnstown and Indiana, Pennsylvania, who planned to eventually use Charlie Bird's "rowboat" to escape to his home in Westsylvania), Jerome Harrison, and some Indians were sent to replace the invalids who had returned to headquarters.

Mr. William C. Wetherill, then a comparatively young man but one who had already given striking evidence of the energy and skill that subsequently made him chief engineer of the Mexican National Railway, was promoted to principal assistant engineer and ordered to assume charge of the corps.

Mr. Wetherill left San Antonio on June 25 but owing to an unavoidable delay in obtaining provisions and transportation, did not reach the Caldeirão do Inferno, seventy-six miles above San Antonio, until July 4. He found the seven men left there by Captain Stiles in a deplorable condition. All of them had suffered from repeated attacks of chills and fever. Mr. Preston himself had been seriously ill for a week, and even some of the new recruits, who had accompanied Mr. Wetherill or arrived two days before him, were almost immediately stricken with the fever. On July 5, so many of the men were unwell that all physical work on the line had to be abandoned, and under a report of the sixth, Mr. Wetherill wrote to Mr. Buchholz, chief engineer in general charge of surveys, *"I regrettably report but three men, McIlvaine, Schele and Young, fit for duty."* This was the last word received from Mr. Wetherill and his corps until they emerged from the forest on August 8 at a point on the south bank of the Jaci Paraná, one half mile below Las Pedro, a camp of rubber gatherers established by Don Pastor Oyola.

As the final location of the railway had been completed from San Antonio to a point near San Carlos, there was no reason for any longer maintaining a full survey corps at headquarters, and, about the time of Mr. Wetherill's departure, it became known that the corps under Mr. C. S. d'Invilliers, would also be moved to some point further south as soon as the expected supply of provisions arrived from Pará. The writer eagerly embraced an offered opportunity to give up the somewhat monotonous routine of duty at headquarters and accept a subordinate position on this corps. Not the slightest intimation had been given to any of us regarding our destination, and the studied reluctance of Mr. Bird on this subject at once created a suspicion that the region we were about to traverse had no characteristics that he could advertise with advantage to the service. It was evident, however, to those who were familiar with the maps in the office that the three parties already in the field covered practically the entire route between San Antonio and the Caldeirão do Inferno except that part passing through the region east of La Concepcion and immediately north of Jaci Paraná.

Our English scouting predecessors, by their representations, had most thoroughly blackened the reputation of the whole country bordering on this river, and their statements were confirmed by a proverb current among our Indian boatmen to the effect that a white man could not stay on the banks of the Jaci Paraná and live, as well as by the experience of our own chief engineer and several men on the steam launch, who, during a hasty exploration of that river, had nearly all suffered from unusually violent attacks of fever.

At 4:00 p.m. on June 27, Mr. Hepburn, with the tug *Juno* and the schooner *D. M. Anthony* in tow, arrived at San Antonio amid universal rejoicing. On July 1, the little river steamer *Villa Bella* brought additional supplies from Pará, and Mr. d'Invilliers at once issued orders to prepare requisitions and have everything in readiness for the departure of his corps on July 9.

A group of invalids gathered under the shade of a palm tree and with feeble voices attempting to sing was the only reminder of our great national holiday.

Nevertheless, several occurrences made the Fourth of July 1878 memorable in the history of the expedition. On the previous day, machinists had finished putting together the parts of the first locomotive ever seen above Pará, in all the country drained by the Amazon. It was a fine specimen of American workmanship and bore conspicuously on its cab the name *Colonel Church*. On the Fourth of July, for the first time, we raised steam on this locomotive, which was made to travel over every piece of track then laid at what would one day become the city of Porto Velho, in the vicinity of San Antonio, while, with ringing bell and shrieking whistle, we raised a veritable pandemonium that could be heard for many miles around.

The performance of Colonel Church's namesake was eminently satisfactory until an attempt was made to run over a piece of side track containing a horseshoe curve that Mr. Collins had personally constructed by memory alone. There, the spectacular exhibition was abruptly and embarrassingly terminated by the engine falling off the rails. It was replaced on the following day, continued in faithful service during the remainder of our stay at San Antonio, and, for aught the writer knows to the contrary, may be there yet, among the rusting ruins that are the only monuments of our ill-starred enterprise.

From the time when the first party reached San Antonio on the *Arary* until July 4, there had been 117 working days. During this time, we had laid two miles of temporary and a half a mile of permanent track, upon which a locomotive and one platform car were then ready to run. For a distance of five miles further, the roadway was being graded. The right-of-way, one hundred feet wide, had been cleared of timber and brush to a point fourteen miles south of San Antonio, and the permanent location had been established and marked on the ground for sixteen miles from the same place. To attain these results, the engineers had been compelled to cut and survey lines through the forest aggregating 108.4 miles in length.

Frequent and severe attacks of fever had seriously impaired the health and usefulness of our chief engineer, Mr. C. M. Bird, and it was generally known that, unless there was great improvement in his physical condition, he must soon sever his connection with the enterprise and seek refuge in a more healthful climate. Nevertheless, it caused great surprise when, on July 15, we read the official announcement that his resignation had been accepted and Mr. C. S. d'Invilliers appointed to succeed him.

Two days later, Mr. Bird, accompanied by John Cox, R. E. Johnson, and A. P. Scull, bade us farewell and embarked on the *Juno* for Pará. The very natural regret, felt by all at parting with the leader who had hitherto guided and directed our movements in the wilderness, was to some extent mitigated by a general feeling of satisfaction over the selection of his successor.

Though Mr. d'Invilliers was then a young man, only twenty-six years of age, he had already manifested in his work such executive ability, hard common sense, and inflexible determination, that he was regarded with confidence and respect by all his associates. After graduation from the Polytechnic College at Philadelphia, he had, from 1870 to 1874, been employed as an assistant engineer in constructing the low-grade division of the Allegheny Valley Railroad. He then entered the service of the Pennsylvania Railroad at Pittsburgh, where he was in charge of important constructive work until 1876, when the requirements of the service caused his transfer to the Philadelphia Division of the same railway. There, he continued to serve with great credit on works of construction and maintenance-of-way until the time of his departure for Brazil.

In addition to technical training and experience, he possessed the equally important qualification of vigorous health, which rendered him capable of more than ordinary physical endurance, and his popularity among the members of the engineer corps secured for him the loyal support of his subordinates at all times.

The promotion of Mr. d'Invilliers and an unusually severe attack of fever of which he was the victim three days later made it necessary to place someone else in charge of Corps No. 1. Who that person would be was unknown until the evening of July 8, when the writer was summoned to the office of the chief engineer, whom he found half sitting, half reclining in a hammock. His flushed face and bloodshot eyes, a large map in his hands, several more on the floor at his feet, some drawing tools, and a bottle of quinine within easy reach, all combined to tell the already familiar story of conflict between disease and duty.

After a few preliminary remarks, Mr. d'Invilliers said, "You must, at least temporarily, take charge of my party. Captain Symmes will manage your transportation. Have everything in the canoe early tomorrow, and when ready, report for instructions."

An inability to obtain any connected and continuous record of events other than contained in his own diary compels the writer, in the six short chapters that follow, to relate only such facts as came to his personal knowledge while in temporary charge of Corps No. 1, and the unavoidably frequent references to himself make it convenient to adopt the form of a strictly personal narrative.

This course has been pursued with extreme reluctance, as the only possible way to avoid a gap that would otherwise exist in the history of the expedition, extending from July 9 to October 11. It has the effect of giving undeserved prominence to a corps that encountered no worse misfortune than that of being placed temporarily under the leadership of one who was at the time acquiring his first practical experience in railway work. The reader will, nevertheless, be able to obtain from these personal recollections some faint, though inadequate, conception of the various obstacles, difficulties, and dangers, which men of greater experience, such as Byers, Wetherill, and Bruce, had to contend against.

XXI. The Canoe Voyage of Corps No. 1 to San Patricio

Some drops of joy, with draughts of ill between,
Some gleams of sunshine "mid renewing storms."
—Burns

The standard railway engineer corps of 1878, even when organized for service in the most salubrious climate, usually consisted of eighteen men, and in the region above San Antonio, where three-fourths of those employed had at times been disabled by sickness, it was absolutely essential to efficiency that each corps should have, not only the usual numerical strength, but additional men to take the place of invalids and a sufficient force of Bolivian Indians to move camp quickly and transport provisions without interference with the fieldwork of the party.

As previously noted, four corps had been formed out of a force of engineers originally intended for three. The ranks of these had been constantly depleted by the necessary details of men for office work and to superintend construction near San Antonio, by the sickness of some, and by the enforced return of others to their homes.

Vacancies in the lower positions could be filled by intelligent laborers, but it was impossible to supply the deficiency in men capable of serving as assistant engineers. We were, therefore, confronted at the onset with the discouraging fact that Corps No. 1 consisted of only twelve men, whose names and positions on the payroll were as follows:

Neville B. Craig	Chief of Party
Joseph S. Ward	Topographer
J. J. Vierra	Draftsman
J. C. Patterson	Rodman
T. C. Maher	Chainman
Frank	Chainman
Walter	Flagman
Alem Bly	Axman
John Ferguson	Axman
Freeman	Axman
Thomas Manning	Axman
Edward Belcher	Cook

Besides our own provisions, camp equipage, instruments, and personal baggage, we were to carry supplies for the corps under Mr. Wetherill, which he had been unable to obtain at the time of his departure.

As only one large canoe, with a crew of ten Indians under a *capitão*, was available for transportation at San Antonio, it was necessary to make two trips between that place and the Falls of Theotonio, where we expected to meet Mr. Buchholz, the principal assistant engineer in general charge of surveys, accompanied by Don Ignacio Arauz and ample facilities for conveying us to our as yet unknown destination.

Before 10:00 a.m. on July 9, I informed the chief engineer that the first division of our party, consisting of Ward, Patterson, Snyder, Bly, Belcher, and myself, under the escort of Captain Symmes, as master of transportation, was ready to embark.

My instructions were then handed to me in a sealed envelope, which I was ordered not to open until we had finally dismissed the San Antonio canoe at the Falls of Theotonio. The obvious reason for this precaution was the fear that if our destination became prematurely known, the notoriously bad reputation of the country we were about to penetrate might cause some fainthearted or fever-stricken individual to return and seriously cripple a corps already too small for the work it was expected to accomplish.

The dimensions of the canoe were approximately: twenty-five feet long with an eight-foot boom and two-foot draft. We were only a few minutes in finding our proper places. Three of the passengers were crowded together in a small space at the bow. Behind them were the ten Indian paddlers, five on each side, attired in palm-leaf hats and coarse white muslin shirts that reached to the knee and were belted in at the waist. The cargo, amounting to a ton or more, was piled high in the middle of the canoe between the two rows of Indians, and behind them was an arched covering made of palm leaves sufficiently large to shelter the remaining passengers in a reclining position from sun and rain.

Standing in the stern behind this shelter and overlooking it was the *capitão*, tiller in hand, giving orders to his men. As we were slowly paddled upstream among the partially or wholly submerged rocks and through the strong eddying currents at the head of the San Antonio rapids, he soon gave evidence that he was not deficient in that skill and dexterity that have made the Indian famous as a navigator throughout South America.

To avoid the strong current, we were obliged to hug the east bank of the river and, of course, encountered many obstructions, such as caving banks, fallen trees, sunken logs, sandbars, and submerged rocks, not to be found in midstream. Our motive power was in some respects superior to steam. In shallow places, the Indians jumped overboard and pushed the canoe through to deep water. In places where the current was too strong for paddling, they ran a line to shallow water near shore and performed the duties usually expected of canal-boat mules on a towpath.

On rare occasions, it was necessary for the Americans to aid in towing the canoe or to render easier the work of the Indians by cutting their own way across some projecting point of land, through the all-but-impenetrable jungle, where piums, a species of sandfly, and mosquitoes stung every particle of human flesh not protected by clothing and where ants, in countless numbers, covered everything, crawling up the legs of our trousers and under the collars and up the sleeves of our shirts in a generally successful effort to arrive at and sting more sensitive parts of the human anatomy than their winged allies could reach.

Familiarity with the torture inevitably accompanying every footstep in the virgin forest did much to moderate our enthusiasm over the magnificent exhibit of tropical trees, plants, and vines constantly submitted for our inspection. The vegetation, extending down to the water's edge on both sides of the river, was so dense that we could not obtain even the faintest idea of topographical features for more than a few yards inland.

Vines, without a leaf, twisted themselves around the huge trunks of trees until they reached the topmost branches, where they burst forth into foliage so luxuriant that it was difficult for the eye to distinguish between tree and vine. Then, to aid in the struggle for sunlight and incidentally increase the complex tangle of vegetation beneath them, these vines would drop hanging trailers from above to take root in the soil below and furnish them with additional sustenance.

Gigantic trees with orchid-covered trunks and buttressed roots, like monarchs of the forest, stood here and there in solitary grandeur, high above the rest, and seemed to look down with contempt upon the silent struggle for existence constantly in progress below and around them.

At intervals, we recognized many solitary trees we had found valuable in constructive work, such as Spanish cedar, lignum vitae, rosewood, and another of immense size that furnished timber closely resembling boxwood. As we proceeded on our way, we saw groves of caoutchouc (*Siphonia elastica*) with increasing frequency, and where the ground was unusually high, we were able to distinguish a few specimens of the castanheira (*Bertholletia excelsa*), which yields the familiar Brazil nut of commerce. We also saw palms of various kinds, but they were not in such abundance near the river as we had found them on the higher elevations in the interior.

Canebrakes were common wherever there was swampy

ground, and these were more difficult to penetrate because of thorny creepers that formed a most intricate network through and over them.

Rich mosses grew on every decaying tree trunk, and tall, graceful ferns, in thick bunches, covered the ground along every water course.

Sometimes, as we quietly rounded a point, we would surprise a flock of ducks feeding or start a panic among a troop of long-tailed monkeys performing gymnastic feats in some overhanging tree. In places, the water edge was lined with a species of worthless waterfowl, which, at a little distance, one might easily mistake for partridge. Every now and then, we were compelled to disturb the solemn reveries of the beautiful snow-white heron standing on one leg in the shallow water near shore. Kingfishers and toucans, perched at a safe altitude above our heads, would inspect us as we passed. Parrots and parakeets were constantly flitting by, and macaws, with gorgeous plumage, flew in pairs above the highest treetops, alarming all other denizens of the forest with their screeching, as we approached.

On shore, the tracks of deer, tapir, jaguars, leopards, peccaries, and other quadrupeds were not infrequent, but these animals, with an almost boundless area of forest to wander through, found no occasion to test the accuracy of our marksmanship by exposing themselves on the riverbank.

Several times, the sharp eyes of our Indians detected what we Americans would never have discovered for ourselves—alligators reposing in the quiet waters of little tributary streams with only eyes and tip of snout above the surface of the water. We had often heard the expression, "He mistook the alligator for a log until he sat on him," but realized then, as never before, that such a blunder was by no means impossible.

The distance from San Antonio to the Falls of Theotonio was only about ten miles by water, but it was 4:00 p.m. when we disembarked below the rapids at the latter place. Then it required the united strength of passengers and crew to drag the boat through the strong current around a point of rocks to a landing place well below the falls. After discharging the cargo, Captain Symmes and the Indians only waited long enough to receive a written report before departing.

The remaining crew had not gone more than a mile before it was necessary to land the passengers while the Indians dragged the canoe through a narrow passage between rocks, where the water was shallow and the current very swift. In one or two other places, the current was too strong for paddling and the Indians had to tow us for short distances. During the day, one of our Americans had a severe chill, followed by fever, and two others, one of them our cook, were far from well.

After a trip of six hours, we landed at San Carlos. The place was nothing more than a small clearing on the east bank of the river, where a palm-thatched shed, similar to the one at Theotonio, had been erected. Our Indians established a separate camp of their own in the forest, while we Americans stored our baggage and hung our hammocks under the shed.

Captain Symmes had frequently stopped at San Carlos before on his way up and down the river but, because of an unexplained mystery with which he was associated, had no love for the place. His pet dog had disappeared one night while there, and he attributed the loss to jaguars, which were undoubtedly abundant in the neighborhood. On a subsequent night, a large tent was carried away from the place under the shed where Captain Symmes had deposited it, and of course, its loss had to be explained.

The corps under Captain Stiles had lived for some time in the forest at San Carlos without discovering the slightest indication that it was inhabited by other human beings than themselves, and in default of a more plausible explanation, Captain Symmes was compelled to charge the jaguar with this second theft as well as the first. The idea was so extremely improbable that the mere mention of it always excited boisterous hilarity in everyone except the captain himself, who chafed greatly under the jests of his companions, doubtless feeling that his long and varied experience of danger by land and sea entitled him to immunity from the implied suspicion that his fears predisposed him to visions of nocturnal prowlers.

Certain of the essential fact that both dog and tent had disappeared, whatever might be the explanation, he never retired at night without swinging his hammock so high under the ridge pole of the shed that he required a ladder to reach it. With a huge revolver by his side, he managed to pass many a night in peaceful slumber, notwithstanding the danger of falling out while asleep, an event which others regarded as quite probable and likely to result in more serious injury than any the hypothetical jaguars were disposed to inflict.

Before leaving Theotonio, the instructions of the chief engineer had been read and the fact ascertained that we were bound for a point on the east bank of the Madeira nearly opposite the famous hacienda of Don Pastor Oyola, generally known as La Concepcion del Morrinhos. From our landing place, we were to survey a line 4.2 miles in length, in a direction S.57° in order to reach the locality through which it had been determined the railroad should pass, there begin a preliminary railway survey, follow a general course S.60° 30' W. to the Jaci Paraná, continue up the north bank of that river until a suitable place for a bridge was found, and finally, effect a junction on the other side of the river with Mr. Wetherill's line from Caldeirão do Inferno.

The enforced delay at San Carlos would not have been unwelcome had it afforded an opportunity to rest and gather strength for the task assigned us, but there was nothing about the place conducive to repose of either mind or body. Nearly all of us were suffering from a skin eruption, similar to but much more irritating than prickly heat, which made it nearly impossible to sleep at night. During the daylight hours, piums hung in clouds about us, and every sting they inflicted raised a minute blood blister, which in two or three days turned black and gave us anything but charming complexions. The most voracious mosquitoes it had ever been our misfortune to encounter were our constant companions by day and night. Mr. Ward attempted to bathe in a nearby stream but had scarcely dropped his clothes before tiny streams of blood were trickling down from a hundred places on his body, where mosquitoes had bitten him and made him look like a criminal fresh from the whipping post.

The Indians, strange to say, seemed to suffer comparatively little from these pests, though protected by only one loose cotton garment from their attacks. I have watched a *capitão* for hours lazily swinging barelegged in his hammock and constantly reaching down between the puffs of a cigarette to remove a mosquito from some part of his body. He seemed generally to proceed in effecting a capture before the insect could indulge in its bloodsucking propensities, but to make sure that none of the vital fluid had been permanently removed from his system, he invariably swallowed the mosquitoes as fast as he caught them.

Our cook continued to have one attack of fever after another and, even in the interval between, was so weak that he could do no work. Though he remained with us for some time afterward, it was only on rare occasions that he could attend to his duties. Maher and Manning were also taken sick with fever while at San Carlos and Mr. Ward with a very much more serious stomach trouble, accompanied by intense agony, which for a time we could not relieve.

During our stay, the Indians returned to Theotonio and brought up all the camp equipage and provisions we had left there, but that only occupied them one of the three days we passed at San Carlos. During the other two, they had absolutely nothing to do, so I frequently availed myself of the opportunity to increase my slight knowledge of their language by conversing with them, generally about the United States in regard to which they always manifested great curiosity.

Thinking that simple, natural phenomena entirely foreign to their own experience would surprise them most, I endeavored to tell them that it was sometimes so cold in the United States that the water in our rivers became solid and would support upon its surface horse and wagons hauling heavy loads. The effect was disappointing, and the expressionless faces of the audience caused me to think that I had failed to make myself understood. Shortly afterward, Mr. Vierra, a native of Brazil, asked me what I had been telling the Indians. On being informed, he replied, "Well, your reputation for truthfulness, so far as they are concerned, is gone. They told me in sorrowful tones that *'Engineer Craig was a good fellow, but a great liar.'*"

Messrs. Buchholz and Arauz returned about sunset on Sunday evening, July 14, and that night, in the little tent, the former related many amusing incidents connected with his travels on the river, some of them mentioned by Mr. Nichols in the account of his reconnaissance with Messrs. Bird, Buchholz, and Arauz.

Mr. Ward had suffered terribly, and nothing that could be done for him seemed to afford more than temporary relief. It was, therefore, deemed advisable that he should remain at San Carlos, under the care of Captain Symmes, until he was able to return to San Antonio, where he could obtain proper medical attention. He had hitherto been noted for uniformly good health when many apparently stronger men were breaking down. His ability, pluck, and capacity for physical endurance made him an exceedingly valuable man on the corps, and his loss seemed likely to prove a staggering blow to its efficiency.

Early on the morning of the fifteenth, the remainder of the party, in two canoes, started on up the river. By noon, we had reached Morrinhos, where it was necessary to partially unload the canoes before our combined force of about thirty Americans and Indians could drag them up the rapids. The transfer of cargo by land required additional time, and it was 4:00 p.m. before we had reembarked above the obstructions in the river. A visitation of chills and fever made the journey far from pleasant to me, but liberal doses of quinine had broken the force of the attack before we stopped for the night at an abandoned *barraca* on the south bank.

At 6:00 a.m. on the following day, we were in the canoes once more. Nothing occurred to break the monotony of the voyage except the shooting of a wild turkey by one of our men. At 11:00 a.m. we reached the point specified in the instructions as the place for commencing our survey.

A little patch of roughly cleared ground on the east side of the river was the only thing to distinguish our landing place from the wilderness surrounding it, and a large island in the river made La Concepcion on the opposite side invisible. The canoe containing Messrs. Buchholz and Arauz had arrived in advance of the others, and the Indian crew was already engaged in erecting the usual palm-thatched shelter for us and our supplies.

The completion of this structure, unloading the canoes, separation of our own stores from those intended for Mr. Wetherill's corps at the Caldeirão do Inferno, and the eating of a hastily prepared lunch kept us all occupied until about 2:00 p.m., when Messrs. Buchholz and Arauz bade us farewell and, with the remaining canoes and all but two of the Indians, continued on their way up the Madeira.

We had failed to notice on maps of South America, among the many places named after saints, the honored name of that distinguished apostle of Ireland, Saint Patrick, and, to atone in some degree for this apparent slight, decided that the initial point of our survey should thereafter be known and recorded as San Patricio.

It was only necessary at this place to erect one tent to cover our dining table, and before dark everything was in perfect order about camp, but our effective working force had already been reduced one-third. The permanent loss of Mr. Ward was discouraging enough, but we now found that Mr. Patterson had been attacked by fever, McKnight was suffering from stomach trouble, and Belcher, the cook, was completely prostrated by the frequent attacks of fever he had previously passed through. For the last, we had to substitute Bly, the best axman on the party and one of those universal geniuses who could fell heavy timer, cook, shoot the head off a duck, run a locomotive, or operate a sawmill better than any specialists P. and T. Collins had in their employ.

We soon found that Sr. Arauz had consulted his own, rather than our interests, in selecting the two Indians left to assist us. One of these was the only man over forty-five years of age I ever saw on the Upper Madeira, *and he died of consumption a few weeks later.*

XXII. San Patricio to the Jaci Paraná

I had rather have a fool to make me merry than experience to make me sad; and travel for it too.
—Shakespeare

Early on the morning of July 17, 1878, we began the survey of our "A" line, a designation used to distinguish it from what we called "X" (experimental) lines. The "A" lines were simply offsets from the river to reach the locality of the projected railway while "X" lines were intended to closely approximate and aid in developing the final railway location.

To carry on, as usually done, the line work, leveling, and topography simultaneously with three independent parties, when much of the time we could scarcely muster men enough to form one, was impossible. Nevertheless, by the evening of July 22, our train party had reached a point 3.4 miles from the river.

Between the seventeenth and twenty-second, one day had been Sunday. Maher had been unwell, though he persisted in the performance of his duties. Ferguson had been attacked by chills and fever but only remained in camp a half a day.

McKnight had suffered from a severe attack of the same disease, and though he lost only one day on the line, he was in a very debilitated condition for some days afterward. Our young Indian followed with repeated and severe attacks of chills and fever at intervals extending over two days. Wilkinson, affected in the same way, was able to be out of his hammock only one day of the five. Belcher, the cook, was never fit to do more than occasionally render some trifling assistance to Bly in camp. Mr. Patterson, originally attached to the party because his well-known vigorous health and strong constitution seemed to preeminently fit him for such service, had been continuously sick in camp with chills and fever during the whole time. I am thus specific, because the facts noted for these few days fairly indicate the sanitary condition of our corps during the whole survey.

On the twenty-first, Mr. d'Invilliers, the chief engineer, accompanied by Don Pastor Oyola; Mr. F. A. Snow, assistant to the resident engineer; and Wyndam Robertson, interpreter, stopped at our camp on their way up the river. Two days later, Mr. Buchholz returned and, finding that our "A" line had reached the top of a terrace formation four miles from the river, ordered us to begin our "X," or preliminary, line from the point already reached.

On the twenty-fourth, with the aid of six or seven Indians furnished by Mr. Buchholz, we began to move camp from San Patricio to a point where a small stream of beautifully clear water crossed our "A" line at a distance of 2.4 miles from the river. Each American carried his own personal baggage, and the Indians, throughout the day, conveyed to the new camp about one-half of our equipage and requisite provisions. Each

Indian carried three loads that day and traveled an aggregate distance of twelve miles.

The next morning, Mr. Buchholz took his departure early and in order to complete the transfer, left three strong Indians in place of the two previously assigned us.

For a long time, the tortures to which we were subjected in Camp No. 2 seemed to pass the ability of human endurance. The forest was infested with wide varieties of flies possessing similar characters and only differing in appearance, one being black and others red, green, and yellow. Our men applied the general name of "sweat flies" to all, because they never attacked in the morning but waited until perspiration had moistened flesh and clothing. Then, by the million, they settled upon us. Through little rents made by briars in our clothing, they would get in close contact with the flesh and there, under our flannel shirts, take their fill until distended to their bloodsucking capacity. They would cover our hands and faces and get into our eyes so persistently that it required two men to keep them out of the eyes of the transitman during the few seconds required to keep their eyes pointed ahead. It is difficult for one who has never had such an experience to realize that these pests could nevertheless, make human existence nearly intolerable and work almost impossible. Every edible on the table was covered with them, but salt, sugar, and syrup were their prime favorites. It was difficult to eat unless at least half a dozen flies accompanied each bite that entered our mouths. The common insects at home can be driven from the table by the fanning use of a brush, but the species we had to contend with on the Upper Madeira were strangers to fear and stuck with such tenacity to every article that appealed to their tastes that each individual one of them had to be separately and forcibly removed. Threats to their well-being and of death had not the slightest effect upon them; so long as permitted to remain in contact with their favorite viands, they would calmly endure with no apparent effort to escape their impending doom. Open a barrel of salt meat, and in a few minutes, swarms of flies made the contents invisible. Let it remain open, and in a short time, all the loose salt, with which the meat was covered, would disappear. Mosquitoes were as annoying as usual. Piums were also abundant but restricted their activity to the daylight hours. For three days, every member of the corps, except the cook, devoted his time exclusively to fighting flies, and for a while, the issue of the conflict seemed doubtful, with chances in favor of the flies.

The campground was completely cleared of vegetation; not even a leaf was permitted to lie within the area we occupied. Large fires were kindled in every unoccupied place in and around camp, as well as inside our closed tents, and by piling on leaves and green wood, such heat and dense volumes of smoke were produced that we were ourselves frequently driven to take refuge in the depths of the forest. Perseverance, in the end, made the locality somewhat less attractive to our persecutors, though we never succeeded in effecting their complete and permanent expulsion.

On the line, of course, we had no adequate remedy, because a smoke dense enough to drive away mosquitoes, piums, and "sweat flies" would also prevent us from doing any work. Ants of many sizes and species were everywhere and on everything throughout the forest. If a wooden transit box were left overnight on the line, it had to be saturated with kerosene to prevent the ants from eating it before morning.

Mr. Patterson was the proud possessor of half a dozen bright-colored silk handkerchiefs, which, worn about his neck, gave him quite a dandified appearance in a wilderness where some of his companions could only boast of one shirt. He stopped one morning on his way to work, washed all six in a little stream that crossed our cut line, and hung them on some bushes to dry, intending to pick them up on his return at night. On the way back, and before we had approached nearer than half a mile to the place where the handkerchiefs were supposed to be, the forest far and near was found to be ornamented with millions of minute pieces of red, white, blue, and yellow silk all in constant motion. Close examination proved that each piece was carried by an ant and not a trace of the handkerchiefs could be found on the bushes.

A species of small red ant was the most numerous and most vicious. They pinched and stung simultaneously, and the effect was similar to that produced by a red-hot needle penetrating the flesh. They occupied every leaf, every blade of grass, every vine, and every tree. Sometimes, the blow of an ax on a tree trunk would make it rain ants on us by the thousands from the foliage above, and there was not a minute passed on the line when some could not be found on any one of us.

Another species of ant often moved in great armies that required as much as an hour's time to pass any given point. Once in motion, nothing but fire or a stream of water could force them to change their line of march. At night, we frequently awoke from a sound sleep to find millions of them walking over us, over our hammocks, over our tents, and over everything in our camp. When the torture of their bites forced us to spring barefooted to the ground, where every square inch of territory was occupied by about one dozen ants, the wildest gymnastic performance followed, while the men were seeking lights, shoes, clothing, and material to fight the advancing hosts of our enemies. As nearly everything was wet with heavy dew and most of the wood in the forest green, it generally required a good little time to build a continuous line of fire entirely across the campground to obstruct their progress. Meanwhile, we had to be careful that the ants did not obtain access to any portable provisions; for one of their armies could easily carry off a bushel of rice in a single night.

Even when mosquitoes, piums, ants, chills, and fevers failed to annoy us, our bodies were always in such an inflamed condition from bites and heat eruptions that restful sleep at night was rarely obtainable. Once, driven nearly insane by almost continuous loss of sleep for a week, I sprang from my hammock at midnight and, in a state of desperation, dashed a quantity of aqua ammonia over my body. It felt like a touché of molten iron but after a brief period of intense agony, brought partial relief.

Brazil nuts were almost the only edible we had found growing in the forest up to this time. They grew upon high trees, a dozen or more encased in a hard shell about the size of a coconut. The monkeys were very fond of the nuts but could not break the external shell and, therefore, resorted to a stratagem in order to obtain the contents. A troop of them threw down the desired quantity of nuts and then hid in the treetop until peccaries came along and cracked the external covering; then, in the twinkling of any eye, the monkeys descended upon them, grabbed all the nuts, and returned to their resting place until the peccaries resumed operations.

In the low ground along our route, it was often a full, hard day's work for four experienced axmen to clear six hundred feet of line for a width of three feet, but on the higher ground of the interior, the forest was more open, and the undergrowth lighter and more easily removed.

One day, at Camp No. 2, it was necessary to send two Indians to our supply station at San Patricio for a barrel of beef. They returned at night empty-handed, explaining that they had been followed by tigers[24] and, being unarmed, had thought it prudent to abandon the undertaking and make a hurried retreat. None of us had as yet seen any of these dreaded animals, and we were not disposed to place implicit confidence in the truth of the story related. On the next day, the Indians were ordered to return with Alem Bly, as an armed escort, carrying a Sharp's cavalry carbine and six cartridges.

About 5:00 p.m., the little party returned, and we all noticed Bly's ashen countenance as he followed the Indians over a huge tree trunk that served as a bridge over a little stream a few yards west of our camp. They had been followed nearly the whole distance from San Patricio by four leopards, which continually circled around them, appearing at intervals on the cut line behind and in front. The Indians again desired to drop the meat, but Bly would not permit them to do so.

---

[24] The Indians applied the word *tigre* indiscriminately to all animals of the feline family.

During the trip, he had fired four times at the leopards, apparently wounding some of them. When he reached camp, he had only two cartridges left—one in his rifle and the other held tightly between his teeth. The incident convinced me that these animals, though calculated to inspire terror on account of their immense size and great agility, were really not dangerous unless rendered desperate by wounds or semistarvation. Not long after, I encountered two of them while going alone through a canebrake penetrated by our preliminary line. They were not more than twenty feet distant when I first saw them, and as I stood almost petrified by fear and armed with no other weapon than a .38 caliber Smith and Wesson revolver, they quietly and most deliberately rose from a recumbent position, yawned, lazily stretched their legs, and walked away.

We had seen many indications that game was abundant in a certain district situated about two thousand feet east of station 36 on our preliminary line, but hunting, as an incident to work, almost invariably resulted in failure. To be successful in killing wild animals where they could so easily keep out of sight in the dense undergrowth, it was necessary for a man to devote his time exclusively to pursuing them at proper hours of the day or night and to familiarize himself thoroughly with their haunts and habits. Tapir tracks, for instance, could be found in the vicinity of every water course, yet, during my whole stay in South America, I never saw one live tapir. In some localities, two good Indian hunters could have kept our table almost constantly supplied with fresh meat, but we had no men to spare for such work. Sickness disabled so many of the corps that instead of prosecuting all branches of work simultaneously, we were often forced to make the same men, on successive days, operate ax, chain, level, and transit, as well as compass in taking topographical notes.

On the twenty-eighth (a Sunday), with Manning, I went in search of a suitable location for our next camp. We both had shotguns and revolvers and, when returning, stopped to explore the game country above referred to. We could only obtain easy access to it, without cutting our way, by following the bed of a little stream that crossed our preliminary line and flowed in an easterly direction at the bottom of a deep ravine. Following this, we soon arrived on the bank of a large creek, where extensive sandbars were exposed on the convex side at every bend. From the tracks in the vicinity, we were led to suppose that it was used as a place for all the tapir, peccaries, and other wild animals in the country to congregate.

In the sand, there were also many round holes where turkeys had been scratching. Manning and I hid in the bushes at points about half a mile apart, and before sunset, I succeeded in killing one large turkey and two guinea fowl. When Manning rejoined me, he insisted in the most positive manner that he had discovered *human footprints* on the bank of the creek below, but the statement made little impression upon me at the time. I regarded Manning as something of an alarmist, and as darkness was rapidly approaching, there was no time to investigate the story. Besides, the question raised as to the existence of savages seemed unimportant, in view of the extreme care they had hitherto taken to keep away from us.

During the two last days of July, five Americans and one Indian were on the sick list, and, in consequence of severe attacks of chills and fever, unfit for service. Returning to camp on the evening of August 1, I found that Mr. Buchholz had been there and had left word that, with additional Indians, he would aid us the next day in moving to Camp No. 3, which we had previously decided to locate near a fine stream of water that crossed our "X," or preliminary, line 1.5 miles from its starting point, 3.1 miles from Camp No. 2, and 5.5 miles from San Patricio. As we had entered upon the dry season of the year, when rains were of rare occurrence, it was thought advisable to diminish the labor of transportation by leaving all our large tents standing in Camp No. 2 and taking with us only the flies belonging to them. These could be used to cover an easily erected and substantial wooden framework, to which our hammocks could be attached. Under the shelter thus formed, we were likely to find adequate protection from any but a heavy and long continued rainfall. One little "A" tent was, however, taken with us to cover our supply of provisions.

It required three days, or until the evening of Sunday, August 4, to convey additional supplies from the river and completely establish ourselves in the new camp. Mr. Buchholz did not remain to see the transfer completed, but before leaving us, he increased the number of our Indians to four. On the first day of his visit, he started shaking. This was followed by high fever, which compelled him to remain several hours in his hammock, and as soon as he had partially recovered, he took his departure with no apparent regret. It was the last time we saw him in Brazil. He resigned soon after and, on August 31, departed San Antonio for Philadelphia. To our great regret, Mr. Patterson accompanied Mr. Buchholz as far as La Concepcion, where it was hoped he would soon regain his health and strength under the care of Don Pastor Oyola. He had been sick almost continuously since landing at San Patricio and had experienced such frequent and severe attacks of fever that it was considered dangerous for him to remain longer with us.

During our first few days in Camp No. 3, there was a decided improvement in the health of the men, probably due to the improved quality and increased quantity of our provisions. Don Pastor Oyola, to whom we were frequently indebted for hospitality and assistance, had sent us some plantains and yuccas. The Indians found some cabbage palm, which we ate with relish. Several times, I was able to kill turkeys, and on more than one occasion, monkey stew served as a most agreeable substitute for salt beef and pork.

One afternoon, about this time, when darkness was rapidly approaching and "the horror-breathing gloom of sunless woods" had settled around our camp, I was bathing in the little stream nearby, when I observed a movement of the dense foliage on the opposite bank and, an instant later, imagined I saw a human face peering at me through the bushes. Grabbing my revolver, which I always kept close at hand, day and night, I fired twice in the direction of the moving foliage and then picked up my clothes and ran to camp.

Really uncertain whether there had been any cause for alarm, I said nothing to the men except that I had seen a movement in the bushes and had fired to scare away any wild animal that might be there. Still, that face, real or imaginary, continued to haunt me, and the impression that it belonged to a creature having a much darker complexion than any of our domesticated Indians clung to me for a long time afterward.

Twice on our way to the Jaci Paraná, we ran into such rugged country that we were forced to spend several days in exploration over a large area, reject parts of our original line, and make changes in our route.

Hardly a man on the corps was perfectly well at any time during our survey, but mere indisposition and general debility were never regarded as sufficient excuses for neglect of duty, and so long as the men were able to keep on their feet, they were always ready for service.

On the morning of August 9, we left one Indian and three Americans sick in camp. While on the line, a new insect plague, not previously encountered, visited us. Swarms of little flying and stinging ants filled the air and nearly blinded the men by getting into their eyes. In the afternoon, four Americans in the same locality and in a very brief period of time were almost simultaneously attacked by violent chills followed by high fever. The effect upon them was such that in five minutes they were as helpless as infants and scarcely able to walk or stand without assistance. This state of affairs made it necessary to stop survey work at a very early hour, as the well men had all they could do to get the invalids into camp before dark.

The next day, we were forced to suspend work and give the members of the corps a chance to recuperate. On the thirteenth, Mr. W. C. Wetherill, accompanied by Mr. Patterson, visited camp in my absence. The former presented a note, from which we learned that Mr. Buchholz was again sick with fever at La Concepcion but, with Mr. Wetherill and Mr. Preston, would leave for San Antonio the next day and that Mr. Wetherill's corps had reached the Jaci Paraná on the eighth and had continued their line one half mile up the south side of that river to Las Pedras, where Don Pastor Oyola had established a *barraca* for the accommodation of his rubber gatherers.

Mr. Patterson, though still unwell and really unfit for service, remained with us.

From the sixteenth to the eighteenth inclusive, we were again engaged in moving camp to a point 3.4 miles farther south, and as two of the Indians were sick, the work proved exceedingly tedious. We were then a distance 4.9 miles from the zero of our preliminary line and 8.9 miles from our starting point on the Madeira.

All the provisions stored at San Patricio had been consumed, and we expected our next supply to reach us on the Jaci Paraná, where we were already overdue. By August 20, all our lard, flour, baking powder, hardtack, and tobacco had disappeared. We still had salt beef, pork, rice, coffee, and sugar, but the men would almost starve before they could be induced to eat salt meat of any kind, and, when our Indians occasionally shot a monkey or found a cabbage palm, there was general rejoicing in camp.

That day, an accidental blow from a machete crippled me so that I was unable to walk and, for four more days afterward, was compelled to remain in camp helplessly watching the rapid depletion of our food supply and the gradual drying up of the only spring of water known to exist within a distance of several miles. Fortunately, on the twenty-third, a heavy rain, the first we had seen since May 2, temporarily increased the flow of water and prolonged the existence of the camp until, on the evening of Saturday, the twenty-fourth, it became evident that our water supply could not possibly last more than two days and our provisions about the same time. The preliminary line had then attained a length of 7.4 miles, and its southern end was 11.4 miles from San Patricio.

It was imperatively necessary that we should reach the Jaci Paraná without further delay, and being unable to move as rapidly as the others, I left camp on Sunday with two Indians and a little "A" tent and passed the night in the forest close to the end of our cut line. I had instructed the others to continue the work on Monday and to follow a route I had proposed to partially cut and blaze through to the river.

Before starting out on the morning of August 26, the two Indians who had been searching for water through the forest came to me and, pointing to the south, exclaimed in evident alarm, *"Caripunas aqui!"* (Caripunas here.) I was aware that we were in territory where those savages were sometimes seen but knew that, while notorious as thieves and murderers, they were the most cowardly tribe on the river and not likely to attack anyone fully prepared to give them a warm reception. By keeping in advance of the Indians with a revolver and double-barreled shotgun while they cut a distinct trail through the underbrush and blazed all the large trees along our route, I succeeded in quieting their fears. Descending gradually, we soon found ourselves in a forest of rubber trees, where marks left by the last flood indicated that the ground was from six to ten feet below extreme high water in the river.

It was evident that the rubber sap had been regularly collected; we could see the scars caused in tapping the trees, and throughout the forest, there existed a most intricate labyrinth of well-beaten paths, designed to reach every rubber tree with a minimum amount of cutting and clearing. These paths were constantly changing in direction and were sure to bewilder any unfortunate stranger who attempted to follow them.

Continuing on one fixed course, for a distance of about two thousand feet farther, we had the satisfaction, at 2:00 p.m., of standing for the first time on the north bank of the Jaci Paraná, and when the transit party arrived three hours later, the Indians had cleared a small area of ground, erected the little "A" tent, and hung our hammocks between flexible saplings and were engaged in a search for something that would allay the pangs of hunger with which we were all afflicted.

We had spent forty-one days between San Patricio and the Jaci Paraná. Of these, six were Sundays, seven were spent in moving camp, and one was lost by the physical disability of practically the entire corps. In the remaining twenty-seven days, actually devoted to fieldwork, we had surveyed the twelve miles from San Patricio to the Jaci Paraná and branch lines aggregating more than an additional mile. Our average rate of progress had, therefore, closely approximated half a mile per day, which, strange as it may seem to those unaccustomed to work in such unhealthy regions and in such dense tropical vegetation, was regarded as entirely satisfactory.

The Jaci Paraná had an average width between the banks of about three hundred feet, but at the time of our arrival, the waterway was in places hardly a hundred and fifty and rarely more than two hundred feet wide. At high water, only the sharp bends in its course would prevent the largest ocean steamships from navigating it for many miles, but in the dry season, there were shallow places, which only large canoes and other craft of equally light draft could pass.

The establishment of Camp No. 5 at the river end of our preliminary line, the removal of our belongings from the old camp, and the solving of the problem of subsistence furnished ample occupation for all on the day following our arrival. Our stock of provisions had been reduced to a moderate quantity of rice, an ample supply of salt, and a quantity of coffee much in excess of our immediate requirements. There were no heavy barrels of beef, pork, and hardtack for our Indians to carry, and this made the transfer from the old camp less laborious than usual.

We expected that provisions would soon reach us from San Antonio but hoped in the meantime to catch enough fish to keep our table well supplied. Failing in that, we would have to endeavor to discover the camp of Mr. Wetherill's corps and obtain such provisions as they could spare.

Early in the day, we found that Don Pastor Oyola had a *barraca* at St. Helena on the river half a mile above us and on the same side as our camp. In it was housed a small colony of rubber gatherers and their wives, from whom we obtained a small quantity of yuccas and plantains. They also had a number of chickens, but these could not be purchased, because the natives were dependent upon them for a future supply of eggs. Early in the day, I got out my tackle, with the idea of catching some of the many fish we could see in the clear water of the river and soon landed a small one, in size and shape resembling our sunfish. In a mechanical way, without paying much attention to the fish, I put my finger in its mouth and extracted the hook. A few minutes later, that finger was bleeding profusely and examination proved that, with a double row of little triangular teeth, each one much sharper than the keenest surgical knife, the fish had taken a slice off the end without my feeling it. I cannot recollect that we ever caught another fish in the Jaci Paraná, though the necessity of replenishing our stock of provisions caused us to try every possible method of capturing them.

Most persons suppose that fish and other animals only become wary as a result of experience with danger, but never was there a greater mistake. One species of fish, particularly abundant in the river, manifested almost incredible discernment and discrimination. They were about one foot in length and could easily be distinguished from other kinds by a spot on each side of their bodies near the tail, which closely resembled the eye of an owl. Dozens would instantly rush to catch insects, worms, or camp offal thrown into the water, but attach the same things as bait to a line, and you might fish all day without a nibble at the hook. With a mosquito bar and barrel hoops, we constructed a set net with a funnel-shaped entrance and baited it with everything obtainable likely to tempt a fish's appetite. By the dozens, they swam around and examined it with the greatest care, but not one would enter. Probably the large number of insects constantly falling into the water from overhanging bushes and trees made it seem to their wise heads unnecessary for them to incur any risk in obtaining food.

However that may be, I have never seen elsewhere fish so wary as in the Jaci Paraná.

The accident to my finger brought to mind an almost forgotten passage in Keller's report, which reads as follows:

It (the dog fish) is not so dangerous to man as the rays or the piranhas, broad fishes of little more than a span's length, which have literally torn to pieces many a daring swimmer. Their two rows of projecting teeth, which are as sharp as needles, are the more to be dreaded, as the terrible creatures are almost always together in hundreds, and they throw themselves upon their victim with the rapidity of lightning, as soon as the water has been dyed with the blood of the first bite, each individual one of the dreadful snapping little jaws tearing off a piece of flesh. Without any doubt these piranhas are a much greater obstacle to bathing than the jacares (crocodiles), whose victims are far less numerous than is generally believed.

In view of the statement just quoted, it is not surprising that bathing in the river never became a favorite pastime with us. Later, when we found that large crocodiles came out of the water every night to occupy a sandbar opposite our camp and that rays could be found along the river's edge almost anywhere, nothing would induce the men to enter the water.

XXIII. The Search for Bruce's Corps

*Complete success alienates man from his fellows,
but suffering makes kinsmen of us all.*

—Hubbard

Our first day on the Jaci Paraná was rapidly drawing to a close; our camp was beginning to assume an orderly appearance, and, notwithstanding our unsatiated craving for tobacco, the shortage of our food supply and the tortures inflicted by the swarms of piums that filled the air about our heads, we were endeavoring to cultivate a spirit of calm resignation to conditions beyond our control, when a little canoe, paddled by Indians and carrying two American passengers, was seen coming slowly up the river. A close inspection of the visitors was required before we recognized the once-familiar faces of "Doc" Lafferty and Edward Green, both of whom had been attached to Mr. Bruce's corps then engaged in carrying the railway survey from a point in the interior east of Morrinhos to a connection with our own preliminary line four miles east of San Patricio. The passengers were only able to mutter a few incoherent words about "fever" and "starvation," hand me a note from the chief engineer, and then permit themselves to be placed in hammocks, where except when administering food and medicine, we left them undisturbed until the next morning. Their physical condition was so pitiable and the change in their appearance since we had last seen them so great that tears came to the eyes of men long unaccustomed to such manifestations of weakness. Mr. d'Invilliers doubtless supposed that his messengers would give me all the detailed information desired, but the men were in a state of collapse, frightfully emaciated and apparently in immediate danger of death. It was useless to attempt getting any facts from them, and not until sometime later did I learn of the long and tedious tramp these fever-stricken and half-starved men had taken from Mr. Bruce's camp to his initial station on the river, of their long wait there without provisions, of their subsequent meeting with the chief engineer, and of his inducing Don Pastor Oyola to send them in a canoe to our camp.

The note from Mr. d'Invilliers contained a statement that Mr. Bruce's provisions were exhausted and his men starving and that they were believed to be near the zero station of our preliminary line but for some reason had been unable to find it and were practically lost. He therefore ordered our whole corps to return as expeditiously as possible to the starting point of our preliminary survey, thoroughly search the forest for Mr. Bruce's corps, and when found, conduct them over our "A" line to San Patricio, where he had deposited a fresh supply of salt beef, hardtack, molasses, and a few other articles for their use.

One thing puzzled me greatly: why Mr. Bruce's corps should be starving so long as they had Indians to carry provisions and a cut line leading directly to the river at San Carlos. The route was a long one, but I could see no reason for starvation so long as it was unobstructed. It was subsequently ascertained that, anticipating no trouble in making a connection with our line, Mr. Bruce had sometime previously directed his provisions to be delivered at San Patricio instead of San Carlos, because the former place was then much nearer to his camp. On August 23, one of the Indians used as a carrier at San Carlos had been killed by savages while cutting wood for a paddle in the forest. This so alarmed all the Indians attached to Mr. Bruce's corps that they could no longer be induced to travel over the route by which they had previously transported his supplies.

Reflecting that our provisions were practically exhausted, that we were dependent upon the rubber gatherers above us for yuccas and plantains, that Lafferty and Green could not be moved and required great care, that considerable time and labor would be involved in moving all our equipage a distance of eight miles, and that having arrived at the northern end of our preliminary line, we would still be obliged to send four miles to San Patricio and draw upon provisions intended for Mr. Bruce, it seemed to me impracticable to use the whole party in the proposed search.

Accordingly, I made preparations for departure the next day and arranged to take with me only Frank Snyder and two Indians. Those remaining in camp were told to expect our return on the evening of the third day and for those three days to subsist upon such provisions as we then had but that if we failed to put in an appearance at the time specified and every other means of obtaining food failed, to shoot as many chickens as they required at the *barraca* above our camp. At an early hour on the morning of the twenty-eighth, we set out.

Besides our revolvers, hammocks, and a few yuccas, Snyder carried a Sharp's carbine and I a double-barrel shotgun. The two Indians followed with a large box of matches, a small tin water bucket, a coffeepot, a small frying pan, some salt and coffee, and a very few yuccas.

At home, such a journey would have been a trifling matter, but over the stumps of trees, vines, and bushes on our roughly cleared line, with one of my legs still stiff from the wound inflicted only eight days before, an average rate of half a mile an hour was the best possible for me to maintain.

We passed that night in our hammocks and with no other shelter than the foliage of the trees to which they were tied. The evening and morning meal put an end to our little supply of yuccas as well as to a ray the Indians had speared while crossing a creek. Thereafter, we were to be dependent upon game for food and, of course, were constantly on the alert to discover it.

About 10:00 a.m. on the twenty-ninth, we were nearing the same deep ravine by which Manning and I had gained access to our hunting ground one month before. As we started to cross, Snyder was a short distance in front of me and the Indians following behind. When Snyder had gone some twenty feet down the slope, he suddenly dropped to his knees and raised his rifle as if to shoot. I waited patiently several minutes, and, as he was still trying to draw a bead on some object below, I endeavored, from my more elevated position, to look over him and see what had attracted his attention. Soon, I caught sight of the bright-red comb of a turkey and let fly with both barrels of my gun right near and close to Snyder's head. He turned with a look of reproach and was about to say something, when an answering volley of shots was heard about three thousand feet to the eastward. No human being who understood the use of firearms had ever penetrated this country before ourselves, and there could not be the slightest doubt that Mr. Bruce's men were signaling to us.

Overjoyed at the early success of our mission, we fired several shots from our revolvers to let Bruce's men know we understood them. Then leading the way to the bottom of the ravine, closely followed by Snyder, I only stopped to pick up an immense turkey gobbler and throw it to the Indians before plunging down the ravine through water and deep black mud to the bank of the creek where I had previously been with Manning. The water was only waist deep, and we were soon on the other side where we waited long enough to fire our revolvers again and catch the direction of the answering shots. Then we plunged into the jungle, and, tearing our way through like madmen, a few minutes later came upon Mr. Bruce, accompanied by several members of his corps, among them Messrs. George W. Creighton, John W. Clark, and Edward N. Stewart. It was evident from their appearance that they had suffered greatly. Mr. Bruce explained that his camp was several miles to the north, that they had been attempting to find our line, and that for several days they had been without provisions, except rice, which their stomachs would not retain when cooked without salt.

On being informed that we had salt, coffee, and a large turkey waiting for them on our line, it required no pressing invitation to secure their presence at a promised feast. I doubt whether any one of the party has ever since looked upon a Thanksgiving turkey without thoughts of that joyful reunion.

The entertainment was not altogether one-sided. We found that some of Mr. Bruce's men had tobacco, and this was quite as gratifying to us as the turkey was to them. When we had finished a hearty meal, enjoyed a smoke, and taken a little rest, the whole party started over our line to San Patricio, where we arrived about 4:00 p.m. Then there was another feast without ceremony, cooking, dishes, or table. Hardtack and molasses were the only articles on the menu, and of both, we consumed a prodigious quantity, simply taking the pieces of hardtack out of one barrel and dipping them into the molasses taken from another.

My next thoughts were of the destitute camp on the Jaci Paraná. We, therefore, fired three shots, the signal agreed upon in case we were attacked by savages, and half an hour later, that prince of good fellows, Don Pastor Oyola, with a boatload of armed Indians was with us. After having our urgent necessities explained to him, he promptly agreed to comply with my request that Snyder and myself be sent back by water with a moderate supply of provisions.

Early on the morning of Friday, August 30, a canoe, manned by a full crew of Indians and laden with yuccas and plantains, was waiting for us at San Patricio. We added to the cargo a small quantity of hardtack and molasses and, after a prolonged and general handshaking, left Bruce and his men to succor the invalids he had left behind, move camp to a point near where we had found him, and complete his survey by making a connection with ours.

Snyder and I were in a very self-satisfied mood as we watched the Indians vigorously plying their paddles, not only because our assigned task had been accomplished, but also for the reason that we felt sure of reaching camp that night and within the three days it had been expected the food on hand would last.

When, however, we reached the mouth of the Jaci Paraná late in the afternoon, the *capitão* ran the canoe ashore near one of Don Pastor's numerous *barracas* and informed us that, on account of business for his patron, we must remain there all night. I had no authority over the crew; it was useless to remonstrate, and I could not explain my reason for haste, but I felt sure that failure to reach camp that night would result in a slaughter of Don Pastor's chickens at San Helena early in the morning and my guilty conscience troubled me sorely.

On many occasions, Don Pastor had played the part of a Good Samaritan to Americans who were sick or in distress and his hospitality was proverbial from the headwaters of the Madeira to the seaboard at Pará. It was maddening to think that, at the very time when his men had been withdrawn from their usual avocations to go on an errand of mercy, the beneficiaries of his benevolence would very probably be engaged in stealing his chickens and that I had authorized the theft. What would he think of me and of such bad ingratitude?

I had never been over the country between our camp and the mouth of the Jaci Paraná and did not know the distance, but the labyrinth of rubber paths were known to extend all the way, and I determined to set out alone on foot. It required two hours of very rapid walking to cover an air line distance of probably a mile and a half. Once on a path, it was only possible to advance or retreat, and there was no telling where its many twists and turns would lead, but endeavoring always to keep as near as possible to the riverbank, soon after sunset, just as the gloom of night was settling over everything and threatening to envelop paths and forest alike in Stygian darkness, I suddenly came out of the forest into the clearing around our camp.

It was evident that none of the men had expected us back at the appointed time, and one or two, whom sickness had unnerved, were inclined to doubt our returning at any time. Except salt and coffee, their provisions were exhausted, and a more dejected crowd I never set eyes upon. My first act was to inform them that provisions would arrive in the morning, and my second was to inquire as to the safety of Don Pastor's chickens. One of the men confessed that he had encountered a solitary chicken at some distance from its home and could not resist the temptation to shoot it. Fortunately, the loss was never attributed to us, and my reputation, in the eyes of our benefactor, remained untarnished to the end.

XXIV. Camp Life in a Rubber Forest

A solemn realm of forest and of flood.
—Bayard Taylor

Only one incident worthy of note had occurred during our absence. Manning had been suffering from that general debility that always follows frequent attacks of chills and fever. Another Irishman of the party, to whom he went for sympathy and advice, told him that what he required was "a good cleaning out," and together, they proceeded to search my medicine chest for something to effect the desired result. As nearly all the medicines bore Latin inscriptions, they were puzzled until they came upon a vial labeled in plain English "Tartar Emetic." The amateur medical practitioner was ignorant of the character of the drug and of the proper amount to take, but with some hazy idea that any emetic would prompt the desired cleansing, assured Manning that it was "just the thing" for a man in his condition and at once dealt him out a liberal dose. It nearly killed the patient who, from that time on, was never able to do a single day's work.

While Snyder and I had been searching for Bruce's corps, our men on the Jaci Paraná became quite intimate with the

Indian colony at St. Helena and were loud in their praises of a beverage manufactured there, which they were very anxious I should try. A little inquiry convinced me that they had been indulging in the far-famed *chicha*. I told them it was made by some Indian women, as Keller says, "not always the youngest or prettiest," who sat around a trough, in which they deposited thoroughly masticated grains of Indian corn; that water was added to the mass, and, after fermentation, was drawn off, filtered, and served to them. The kindhearted Indians at St. Helena were so careful in regard to the neatness of their apparel and so fastidiously clean about their persons and in their house that few of our men paid any attention to my explanation. In passing the *barraca* the next day, some of these unbelievers were shocked at an assurance from the Indian women that the preparation of their favorite beverage would be expedited if they would enter and *help them chew*.

As our camp was in the heart of the district that produced the best quality of rubber shipped to Pará, it may not be out of place to say a few words here in regard to the manner of collecting and curing this valuable product. It is astonishing how many different trees in the Brazilian forests have apparently the same milk-white sticky sap as the *Siphonia elastica*. The tree has a general resemblance in size, bark, and leaves to our elm. It is unlike most other trees along the Madeira in the fact that it is found in extensive groves and flourishes best on ground from two to ten feet below flood level in the rivers.

It is, therefore, only during the drier months of the year that the sap can be collected. For many miles, both sides of the Jaci Paraná had a fringe of rubber trees extending an average distance of about two thousand feet inland. This territory, so far as Don Pastor Oyola had taken possession of it, was divided into districts, each having its *barraca* and proper quota of Bolivian Indians, male and female, all in the charge of a *mayordomo*, who was usually a man of pleasing manners, gentlemanly appearance, and considerable intelligence.

Each district had a certain number of *estradas* or paths, leading to as many rubber trees as one person could attend to in a day. At daylight, the trees were tapped by making two or three incisions in the bark with a diminutive hatchet, which had a cutting edge of only one inch in length. Below each cut, a tin cup was fastened to the tree by pressing a sharp point on its raw edge into the bark, and a little wet clay was then applied below each incision to direct the flow of sap into the cup. About 11:00 a.m., the sap would cease to flow, and the Indian women would then collect and carry it to the *barraca*, where, by an extremely simple process, it was reduced to crude rubber during the afternoon.

Over a fire of palm nuts, a large earthenware funnel was placed to concentrate the fumes. A wooden paddle was then dipped into the sap and given a few turns in the smoke made by the palm nuts, which caused all the adhering liquid to coagulate. This process was repeated until a large piece of rubber was formed around the paddle, which was then removed by cutting. The drippings at the end of the day, mixed with leaves and dirt, were gathered up and made into lumps of irregular shape, which, on account of their inferior quality, were sold at a lower price in the market at Pará, under the name of "Negro heads." While the pieces of rubber were still soft, the initials of the *seringueiro* were stamped upon them. At first, the rubber is pure white, but after a few days' exposure to the sun, it acquires the color found in the commercial product.

Don Pastor Oyola was said to collect sixty-four thousand pounds a year, which, at current prices in Pará, would yield him a gross annual income of twenty thousand dollars.

We had no canoe, nor could we find one anywhere on the river, so on Saturday, August 31, with Snyder, I went on foot, over winding rubber paths, to find the camp of Mr. Wetherill's corps and, after a long tramp, discovered it on the opposite side of the river about half a mile below Las Pedras (The Rocks), as the upper *barraca* of Don Pastor on the Jaci Paraná was called. The air line distance between our camps was about two miles, but by the circuitous route we were compelled to follow, the distance was nearly doubled. Mr. George A. Schele had been left in charge, and his men carried us across the river on a raft. We only remained long enough to ascertain that Schele could furnish us with a few days' supply of flour and salt meat, which was conveyed to us on September 4 by the chief engineer on one of his periodic tours of inspection to the various camps.

Mr. Wetherill accompanied Mr. d'Invilliers on this trip, and on their return from Las Pedras, they brought with them Rodman McIlvaine and Athur P. Herbert, both unfitted by their physical condition for further service.

Having arranged for a temporary supply of food, we then proceeded with our survey of the river, and for a few days, all went well, despite the usual amount of sickness in camp and the usual tortures inflicted by various insects, among which piums were by far the worst. They filled the air about our camp all day, made existence a burden, and forced us to do all clerical work and drafting by lamplight.

XXV. Our Anthropophagous Acquaintances

The very tigers from their delves
Look out and let them pass, as things
Untam'd and fearless like themselves.
—Lalla Rookh

On September 7, I received, by Indian messenger in a canoe, a letter, written by the chief engineer at La Concepcion two days before and containing the startling intelligence that John King, the cook on Bruce's corps, had been killed by savages. The letter also disclosed the fact that the tragedy had occurred close to the place where Manning had claimed to have discovered human footprints in the sand and not far from the point where Snyder and I had found Bruce and his men.

Mr. d'Invilliers, after leaving our camp on the fourth with Wetherill, McIlvaine, and Herbert, had stopped that night at La Concepcion, where he met Captain Symmes. Leaving the others to recuperate, on the morning of the fifth, he took Captain Symmes and a few Indians to carry provisions to Mr. Bruce's corps. The little party crossed the Madeira in a canoe to San Patricio, where they found Mr. Samuel Hoff, transitman on Bruce's corps, who had come there on a foraging expedition, doubtless intending to signal Don Pastor and obtain some edibles from him, as he had seen Snyder and me do a few days before. Finding that Mr. d'Invilliers had the requisite supply, Hoff joined him and Captain Symmes on the five-mile trip over our "A" line to Bruce's camp, where they arrived while the Indians, burdened with provisions, were still far behind.

Hoff was greatly surprised to find none of Bruce's men in the vicinity, because it had been expected that the survey would be completed that day in time to have the whole corps lunch in camp. His astonishment was still further increased a few minutes later when he discovered that all the hammocks, supplies of various kinds, and much of the men's personal baggage had been removed. Finally, attracted by the smoke of a fire used by the cook, they went to it and there found the dead body of John King pierced by three long arrows.

The three Americans naturally supposed that Bruce's party had been driven from camp by the savages some hours before and had been forced to take refuge in the forest. They regarded it as dangerous to remain where, under cover of the dense vegetation, a large band of the savages could easily surround and murder them. They thoroughly appreciated the fact that firearms were as useless as popguns to protect them from the attacks of these bloodthirsty barbarians, who could glide as noiselessly as shadows through the thickest jungle and, undiscovered, approach near enough to spear or club their victims. Their force of three men seemed inadequate to attempt searching the forest for the absent corps, which had apparently been routed by superior numbers. Furthermore, it was feared that, if the unarmed Bolivian Indians learned of the murder, they might abandon the provisions, run for the canoe at San Patricio, and leave the Americans without means of crossing the river.

They hastened, therefore, to intercept the Indians, return with them to La Concepcion, and there obtain a sufficient force to scour the forest and rescue Bruce and his men.

Just after passing our Camp No. 2, where the tents had been left standing, Mr. d'Invilliers looked back and saw that the Indians, for some unknown reason, had started a fire among the leaves close to the tents. Supposing they intended to stop there for lunch, he told them not to do so. They replied with very significant shrugs, "It is no good place to stop." They had evidently discovered something to excite their alarm.

Reaching San Patricio about 2:00 p.m., Mr. d'Invilliers was preparing to cross the river, when unusual noises in the forest caused him and his little party to seek shelter behind trees, where they prepared to resist an immediate attack from the savages. The noises increased, and a few minutes later, Bruce and his men came down our cut line on a run and made a rush for the landing.

They had returned to camp shortly after the departure of Mr. d'Invilliers, discovered King's dead body, and convinced by the amount of plunder carried away that a large force of the savages was in the vicinity, had taken an unceremonious leave. Indeed, it was hardly possible to remain, as the marauders had left them without food to eat or hammocks to sleep in. Along their entire route from camp to river, they had been constantly apprehensive of attack. The fire started by Mr. d'Invilliers's Indians at Camp No. 2 had ignited four tents, and these were all blazing when Bruce arrived. Nothing more was needed to accelerate their pace. The fires seemed to furnish positive proof that the savages were ahead of them and likely to cut off their retreat to the river. It required a number of trips with the canoe to transport the entire party to La Concepcion, and Mr. d'Invilliers was the last to embark.

The next day, an armed party of ten Americans and ten Indians revisited Bruce's camp, buried King's body, took possession of the camp equipage, and withdrew to La Concepcion. Mr. Bruce had, fortunately, about completed his survey by making the connection with our line, so there was no reason why he should remain longer in the vicinity, but this very connection had opened up to the savages a cut path leading directly from his camp to ours, only eight miles distant.

It was not strange, therefore, that the chief engineer had thought it necessary to give us early information in regard to the plundering of Bruce's camp and the fate of his cook. The fact that we had passed unmolested through the region where Bruce's camp was located, long before he had reached it, seemed to us significant. The depredation must have been committed by some nonresident savages. These could not have been Caripunas, who lived south of the Jaci Paraná, because they would have encountered our party first, while moving to the north. Some years before, according to Keller, the Englishmen at San Antonio had been attacked by the Parentintins, and the only tenable theory seemed to be that the same tribe had followed along the cut lines of our preliminary survey, then connected and continuous from San Antonio, to Bruce's last camp.

This would explain the disappearance of Captain Symmes's dog, the stealing of his tent, and the subsequent murder of the Bolivian Indian at San Carlos, where at an earlier date, Captain Stiles had seen no indication of the existence of savages. Assuming this theory to be correct, our own camp was likely to be the next objective point on the route to the Parentintins and some of us probably the next victims.

Keller's report was then the Baedeker of the Upper Madeira, and the information we were able to glean from his pages regarding the Parentintins was far from reassuring. In one place, he spoke of them as "The ill-famed Parentintins, anthropophagous hordes, always ready for robbery and murder." Referring to their attacks upon seringueiros, he said, "The chances of being murdered and roasted are heavy odds against the acquisition of a few pounds of India rubber." He told "of a whole family having been murdered and roasted by them." Further on, we were informed that "The tibiae of victims are used as flutes." He finally dismissed the subject in the following paragraph: "We know, at any rate, what we have to expect from them; and, even if the certainty be not always a very comforting one, it spares us the pang of disappointment, and enables us to prepare for all eventualities."

This passage was doubtless intended to amuse readers living in the latitude of Carlsrhue, where it was written, but failed to increase the buoyancy of our spirits on the Jaci Paraná.

We had a miscellaneous collection of firearms, but the supply of ammunition for them was nearly exhausted. Really no effective defense was possible against savages, who realized the great advantage they had over us, both in facilities for stealthy approach and the superiority, at close quarters, of their noiseless weapons. Our one source of confidence was the fact that the Parentintins were ignorant of the use of firearms and regarded them with superstitious awe. Though a number of rifles were in Bruce's camp when it was plundered, not one was touched. Our Bolivian Indians told us that it was a very unusual thing for the savages to attack a man carrying a gun or rifle and that occasional fights with seringueiros had taught them to dread such weapons, which they believed, in some mysterious way, gave their opponents command of the thunder and lightning. We used, nevertheless, every possible means to protect ourselves against attack. It would not have been surprising had a realization of our perilous position created a panic, but the superabundant privations and tortures, incident to everyday life on the Jaci Paraná, preoccupied our minds and diverted our thoughts from the contemplation of such uncertain afflictions as the future might have in store for us. There were occasions, however, when unearthly sounds in the forests at night caused us to jump from our hammocks in alarm or, during our troubled sleep, to imagine that we heard the strains of barbaric music produced by savages with instruments made of our own shinbones.

Mr. George A. Schele, then in temporary charge of the few men remaining on Mr. Wetherills's corps near Las Pedras, was a native of Sweden and, like some others, was disposed to regard with satisfaction the rapid promotion the sickness of more experienced men made possible. Placed for the first time in charge of a field party, Schele had on several occasions expressed a determination "to make a record" for himself and, with this end in view, was working his fever-stricken and half-fed men early and late to the limit of their endurance. The same warning, sent by the chief engineer to us, was conveyed to Schele but for several very good reasons, failed to alarm him.

Though Caripunas had in previous years been seen on Schele's side of the Jaci Paraná, the Parentintins had never been known to cross that river. The stream itself, which was unusually wide at Schele's camp, with its crocodiles and biting fish, afforded him some protection, and no direct pathway had then been opened to connect Wetherill's line with ours.

Schele argued that, in any event, our camp would naturally be the first one attacked, so he felt perfectly secure. His men were not, however, in sympathy with his desire to make a record. Schele's attention was from time to time directed to faint trails through the forest, which they themselves had made. Unusual sounds, emanating from places in close proximity to camp, disturbed his sleep at night and occasionally pieces of wood were hurled with great violence against his tent. Finally, two of Schele's men, Rulon Gray and James T. Young, returned to camp, apparently in great alarm, with the report that they had caught glimpses of Caripunas attempting to approach them from the forest. They vowed that they were perfectly willing to take the chances of a fair fight but would not sacrifice their lives in a foolhardy contest with an invisible foe.

That quieted Schele's ambition to make a record and terminated his work. Until removed to San Antonio, he and his men could always be found in camp.

On his return trip from La Concepcion to San Antonio, with Bruce and his men, the chief engineer stopped at San Carlos and, from that place, walked several miles inland to Byers's camp in order to warn him, as well as the others, of the impending danger from Parentintins. Reaching the camp about sunset, he presented Byers with a quart bottle of French brandy, probably intended as a nerve tonic to strengthen that gentleman for the prospective tales of savage atrocity about to be inflicted upon him. Not wishing to alarm the men, Mr. d'Invilliers deemed it wise to wait until they left camp in the morning before mentioning the object of his visit to Byers.

To those who never knew "Joe" Byers, as he was affectionately called (behind his back) by his men, it may be necessary to explain that his whole mind and heart were concentrated upon his work. There was no possibility of conversing with him long on any other subject than railroads. It was railroad morning, noon, and night; railroad on Sunday as well as during the week; railroad at the table; and railroad in his dreams by night. Though kindhearted, he had a supreme contempt for those who permitted any obstacle, difficulty, or danger to interfere with the accomplishment of the one object he deemed worth living for—the construction of railroads. Naturally, no human being could long endure the strain of working with such unceasing energy and such unvarying fixedness of purpose, unless something occurred to relieve the high nervous tension he was under. This necessary relaxation "Joe" Byers endeavored to obtain in a liquid form, but only on rare occasions and after long-continued, arduous labor. The effect of his potations was only noticeable on account of an unwonted hilarity of manner, and however much the keenness of his intellect might be impaired regarding other matters, on the subject of railroads and railroad work, his mind was at all times perfectly clear.

Unfortunately, Mr. d'Invilliers and his brandy arrived when Byers had long been leading a life of forced abstinence from intoxicating liquors, and when, the next morning, the time came for the chief engineer to deliver his warning and depart, Byers was in such a jovial mood that no subject, not strictly pertaining to railroad work, made any impression upon him. With a face as long and serious as that of a funeral director, the chief engineer gave all the details of the plundering of Bruce's camp and the murder of King by the Parentintins, at the same time mentioning the precautions Byers should adopt to protect his men. Then he waited to see what effect his words had produced, but a respectfully quizzical look from his auditor was the only indication of intelligence. He went over the whole story and his reasons for apprehending an attack by the Parentintins on Byers's camp a second time but with the same result.

Finally, thinking Byers's lack of appreciation might be due to want of familiarity with Parentintin history, he gave him all the facts in regard to the tribe, which he had obtained from Keller's report, infrequently repeating the word *Parentintin*. Then Byers's face began to assume a weary expression, and as Mr. d'Invilliers paused for reply, he said in tones of peevish and fretful remonstrance, "Don—I don't care a d--n for your *pair of two tin canteens.*"

A few weeks later, while Byers was moving camp and his party was divided between the old camp and the new, the savages sacked the former and carried off in the night, while the men were asleep, such articles as they fancied. Among them, I recall the spade used in setting posts, the target belonging to the level rod, and some of the highly polished brass attachments of the transit and level. Fortunately, the midnight visitation was not attended by loss of life, but the men were nevertheless badly frightened in the morning. It was afterward noticed that none of the bodies of men killed by the savages were subsequently mutilated. It is, therefore, probable that both Keller and we were wrong in attributing such depredations as occurred to the Parentintins. They were probably committed by the Acanga Pirangas (Red Heads), who were supposed to live on the Jamary River south and east of San Antonio and were in the habit of covering their heads with some red pigment.

XXVI. Last Days on the Jaci Paraná

The qualms or raptures of your blood
Rise in proportion to your food,
Your stomach makes your fabric roll,
Just as the bias rules the bowl.

It became evident soon after our arrival on the Jaci Paraná that something was wrong at San Antonio. At no time did we have provisions fit for men to subsist upon in such a climate or so packed as to be easily transported by Indians over our cut lines. While in the forest, at a considerable distance from our base of supplies and with no means of making known our wants at headquarters, we cheerfully endured privations, that were to some extent unavoidable, but after reaching the river, where canoes could be sent directly to our camp, we all felt that, however unsuitable in quality, there was no sufficient excuse for any deficiency in the quantity of our food supply.

Instead of improving, our condition on the Jaci Paraná became steadily worse. The small quantity of provisions the chief engineer had been able to furnish was soon exhausted. A larger supply was promised but failed to arrive at the appointed time, and we soon became dependent upon the generosity of the Indians at St. Helena and Las Pedras for yuccas and plantains. We did not have a cent of money to pay for anything. Schele was able to spare a little salt meat and flour, but all these things had to be carried overland, as no canoe was obtainable. Some of our Indians were sick, and the murder of King had so alarmed them that they would not go any distance from camp without an armed escort. The result was that the Americans had to carry their own provisions. To have used our Indians would have involved sending all the Americans fit to travel for their protection.

On September 12, Snyder's condition became so serious that he was taken in a canoe by Don Pastor's men to La Concepcion for recuperation. After the eighteenth, work was practically suspended. On the nineteenth, McKnight and I were the only men in camp not prostrated by fever, and as there was nothing on hand with which to feed the invalids, we had to visit Las Pedras and carry on our own backs as much of a load as we could stagger under. The generosity of the Indian rubber gatherers was taxed to such an extent that it became increasingly difficult for them to supply our wants without impoverishing themselves. At last, we resorted to trading empty tin cans and every article we could spare for yuccas and plantains.

About this time, one man on the corps took the trouble to carefully estimate the cost of our subsistence to the contractors and the result of his calculation was *six cents per diem for each man*. We did not dare to expend recklessly our small supply of ammunition on game, while there was danger of being attacked by the Parentintins, and at the very time when almost every American and Indian in our camp most urgently required it, our supply of quinine gave out. For weeks, the men had been without tobacco; during their convalescence, they keenly felt the deprivation, which made them extremely irritable.

As the rain continued and it became evident that no more work could be done that day, we began to discuss the possibility of obtaining a supply of Bolivian tobacco, the finest the world produces, from the colony that seemed to exist on the other side. Finally, McKnight dared to cross the river. He invited me to go with him, and I decided to make the attempt. The others were instructed, if we succeeded in crossing, to rejoin us at the same place on the next morning, and covered by the rifles of the two best shots in the camp, we plunged in and swam across, wearing all our clothing as a protection against the biting fish. While in the stream, a sunken log, which I thought was a crocodile, came in violent contact with my stomach, and for a minute caused me to regret, more than ever before since, my fondness for tobacco.

From the opposite bank, a Bolivian girl had been an interested spectator of the whole performance and as soon as we emerged from her side of the river, with an unassuming grace and courtesy that would have been a credit to a duchess, conducted us a short distance on a well-beaten path through the forest to a rubber gatherer's camp, consisting of a small area of cleared ground, in which had been erected several thatched sheds.

What little I then knew of the language of the country was limited to Portuguese, which I had acquired during the trip from Philadelphia. On the Madeira, however, Spanish was the language commonly spoken, but this was so corrupted by the Bolivian Indians that even Mr. Vierra, a native of Brazil perfectly familiar with both languages, failed to understand them. Consequently, our conversation was restricted to words we had picked up in camp, aided by signs and gestures. We soon learned, however, that the name of the place was Natividad and that of our hostess Domatila Galves.

In our dripping clothes, we did not make a very attractive appearance. There seemed to be no one about the camp except the Bolivian girl, who conducted us on a central *barraca* about thirty feet square, open on all sides. Underneath it, many hammocks were swinging while numerous guns and rifles and various implements showed that a party of men were employed in the vicinity. The young lady at once brought freshly laundered snow-white suits of cotton duck and slippers. McKnight and I looked at each other in perplexity for a minute, knowing that to disrobe in the forest meant inconceivable torture from ants and mosquitoes, but there seemed to be no other way to make the necessary change. We started to follow one of the paths leading to the interior. The Bolivian girl returned at this point, brought us back to the *barraca*, and insisted that we should make the change there. Then, she seated herself in a hammock and, with perfect composure, turned her back until we were fully arrayed in costumes of rubber-gathering *mayordomos*.

Such charming manners, dignity, and perfect self-control in the heart of a tropical wilderness excited our unbounded admiration, and we did not hesitate to voice the opinion that many a young lady, in fashionable society at home, could with profit have taken lessons in deportment from this unspoiled daughter of nature. Apparently unconscious of the fact that she possessed personal attractions of a high order, attention to our every want, courteous and kind in her bearing toward us, there was yet not the faintest trace of a disposition to encourage undue familiarity.

Before dark, the *mayordomo* with ten Indians returned, and all united in an effort to entertain us and make our stay as pleasant as possible. It surprised us beyond measure to see the young and good-looking head of the establishment step out of the forest so faultlessly attired that his costume would not have been deemed inappropriate on the beach at one of our seashore resorts.

They gave us the best meal we had enjoyed since leaving Pará and furnished us with native rum and French brandy to wash it down, and not satisfied with my skill in rolling cigarettes, the Bolivian girl insisted upon doing it for me.

We had noticed that several of the Indians had been put to work at weaving four palm-leafed mats and wondered what they were for. The mystery was solved when the hour came to retire. Then the four mats were laid, one over the other, directly under the young lady's hammock, which was elevated about two feet above them. There, McKnight and I were told to rest for the night. I fear the girl did not sleep much for, until a late hour, we continued to indulge in rhapsodies over the hospitality we had received and the charms of our Bolivian hostess. It did not escape our attention that although honored by being permitted to sleep in such close proximity to the only lady in the establishment, she was not left entirely unprotected. In her hair were tied by their antennae several large Brazilian fireflies, which, at the slightest motion of her body, would spread their winds and dispel the surrounding darkness, so that the whole hammock area of sleepers under it could be distinctly seen by the entire party. After breakfast in the morning, with a liberal supply of tobacco, we were carried across the river in a canoe and resumed our work, envied by all our companions.

Soon after this, McKnight was stricken with fever, resulting from an eight-mile trip after provisions. Manning became worse and worse until he was entirely unfit for work of any kind. He was generally left in camp as a guard for the cook. He had a ten-dollar shotgun purchased at the store maintained by the contractors in San Antonio and, having expended his supply of shot, kept the gun loaded with small slugs of lead.

One day, in a fit of extreme dejection, he wandered away from camp to escape from the piums, which made the daylight hours there nearly intolerable and, following one of the many rubber paths, finally seated himself on a log. His gun was on the ground at his feet, his elbows rested on his knees, his head was supported in his hands, and extreme weariness and weakness soon made him oblivious to everything around him. Suddenly, he heard a slight movement in the leaves directly in front of him and, languidly raising his eyes, found a leopard cat staring him in the face not three feet distant. Like most Irishmen, Manning had an instinctive preference for a club as a means of defense at close quarters, so, grasping his gun by the muzzle, he brought down the butt with such force on the brute's head as to break the stock in several pieces. The victim did not delay his departure unnecessarily, and the next day, Manning found occupation in laboriously tying together with twine the pieces of his gunstock, but ever afterward, when that gun was fired, the stock and locks would fall to pieces like the famous one-horse chaise.

Manning soon became perfectly crazy, but in one respect, his affliction was peculiar. He seemed to be mentally well balanced during the day, but as soon as darkness came on, his mind began to wander. He imagined that everyone in San Antonio had deserted us and gone home. On one occasion, he seized a rifle and attempted to shoot indiscriminately about camp, saying, *"The devils are after me. I kicked them out once, but they are coomin' with their frens to lick me."* From that time, firearms had to be kept out of his reach at night, but I requested the men never, during the day, tell him of his hallucinations.

Under the adverse conditions described, with an effective force rarely exceeding four Americans and half as many Indians, we continued to search for provisions, look after the sick, and make some pretense of doing work until September 30, when the chief engineer and Captain Symmes arrived with two canoes, containing such a supply of provisions as none of us had ever seen before in camp or even in the hospital. For the first time, we had American canned vegetables and meats. Five of our men were sent to San Antonio as invalids unfit for service, and with them went Mr. Schele and one of his men.

Our original party of twelve Americans, now reduced to five, was reinforced by three men taken from Mr. Wetherill's corps. The survey was soon completed, with the assistance and under the personal direction of the chief engineer.

On October 8, we abandoned our camp on the Jaci Paraná and, after stops at San Patricio and San Carlos, reached San Antonio three days later. Four of the party, including the chief engineer and myself, had severe attacks of fever on the return voyage.

Though I had, in previous years, done other kinds of civil engineering work, the experience narrated must be regarded as constituting the initiation of a tenderfoot into the rough-and-tumble life of a railroad engineer of that period. Furthermore, the histories of other corps engaged upon the work, could they be written, would furnish many incidents of more thrilling interest than any I have mentioned.

XXVII. Disintegration and Collapse

'Tis hard to toil, when toil is almost vain,
In barren ways;
'Tis hard to sow, and never garner grain,
In harvest days.
—Father A. J. Ryan

Landing at San Antonio, October 11, 1878, we found that the primitive structure formerly used as headquarters by the engineers had been abandoned. The new office building had been completed sometime before and was then occupied by engineers, contractors, and clerks.

Though entirely destitute of ornamental features, it was substantially constructed, admirably adapted to the climate, conspicuous on account of its location, and far superior to any similar structure on the rivers above Pará. Offices equipped with every essential convenience for drafting and clerical work occupied the entire first floor. The second story was divided into comfortable sleeping apartments, some of them furnished with American beds and all opening on spacious verandas, where the occupants could assemble for social intercourse, obtain an excellent view of the river and town, recline in their hammocks, and exchange stories of toil and suffering in camp and on the line.

The six months' term of service, to which all employees were bound by contract, had expired on August 19, and, as men could not be expected to continue indefinitely in a service that brought no reward, present or prospective, save privation, sickness, and death, it was not surprising that an exodus had begun at that date, but to their credit, it can be said that only a very small and insignificant minority of the whole number who returned to the States abandoned the enterprise before they were physically unfit to remain.

Few had sufficient money due them to pay for their return passage, and the contractors were under no obligation whatever to send their men home until they had served two years. Departing employees were supplied with orders on Philadelphia and Pará for the amounts due them, but as a general rule, payment was refused when these orders were presented; not until years afterward were many of them even partially paid. The regular steamers usually required cash in payment of passage, and the great majority of homeward-bound passengers were compelled to accept cheap transportation and inadequate food, which the contractors obtained for them on returning lighters and schooners as far as Pará.

Some, unable to afford even the luxury of semistarvation and two weeks of broiling sun on board the lighters, resorted to canoes and rafts. To the present time, no one knows how many of them ever reached the seaboard alive, but many who succeeded found the situation at Pará even more intolerable than that at San Antonio.

The engineer corps, originally consisting of fifty-four men, but subsequently increased to fifty-seven, when we returned on October 11, could only muster twenty-six, and some of these were unfit for duty. It was obvious that, after deducting the number required for office work and construction, the remaining force would hardly be sufficient to form one field party. Mr. Collins had, therefore, made a contract with Joseph Byers and C. H. Patterson to establish the final location of the railway, as far as the Caldeirão do Inferno, at a cost of five hundred dollars per mile.

During our absence, the colony at San Antonio had passed through the unhealthiest season of the year, and thirty-five men whom we left on July 9 alive and enjoying as good health as any of ourselves, three months later were lying in the "banana patch," as the graveyard was commonly called. Of these deaths, seventeen had occurred in the month of August alone, but the number of cases of sickness seems to have reached a maximum in July when the official report of T. M. Fetterman, the druggist, showed an average of one hundred prescriptions a day and from thirty to fifty dispensary patients. On the same authority, it can be stated that as high as three hundred persons were at one time under treatment at their places of residence or in the hospital. Of course, that does not tell the whole story, nor does it give any idea of how many were dosing themselves with quinine without a doctor's prescription.

No record of events that would make it possible to present a complete account of what occurred at San Antonio while we were absent is now known to exist, but some extracts from private letters and other papers submitted to the writer will enable the reader to catch an occasional glimpse of the situation then and afterward.

On July 12, Commander Selfridge of the *Enterprise*, with one of his officers, visited San Antonio, and turning to his published report, we find the following:

> San Antonio is notoriously unhealthy. ... I have never in my life seen a more unhappy and unhealthy body of men than the workmen on the railroad. ... Hardly a single one had escaped attacks of fever and the pale cadaverous looks of nearly all of them were truly pitiable.

In the sworn testimony of Resident Engineer Nichols, given some months later in the English courts, it was stated that: "On July 16, 1878, we first received positive news at San Antonio that the trustees of the trust fund had refused to pay certificates. We had rumors to that effect earlier." This was a crushing blow to the contractors and to all employed by them. The actual cost of the work had greatly exceeded the amounts called for by the certificates of the resident engineer, but even had the initial work been profitable and payment prompt, the delay, caused by sending accounts from San Antonio to England and remitting funds from there to San Antonio through McCalmont Bros. and Co. and Franklin B. Gowen would have kept the railway company at all times heavily in debt to the contractors.

To find that even the inadequate amount then due would not be paid without delay was discouraging and disheartening. The news seriously impaired the credit of P. and T. Collins at home and completely destroyed all confidence in the financial stability of the firm at Pará. The effect at San Antonio appears in the following extract from a letter from Mr. Nichols to Colonel Church, under the date of July 17, 1878. "We cannot have more than 50 men at work now and these are discouraged because they have not been paid."

Mr. Thomas Collins was not, however, one to bend easily to the storm. Ever inclined to:

> Smile at the doubtful tide of Fate,
> And scorn alike her friendship and her hate …

we find him, at the very time Mr. Nichols was penning the lines quoted here, signing a contract with Francisco Mareno, by which Mareno was bound to furnish one hundred men within twenty-four days and one hundred more thirty days later. In this contract, it was stipulated that Mareno should transport his men to San Antonio and furnish them with food, lodging, and medical attendance. For this, he was to receive $1.50 per man for each actual day's work of nine and a half hours.

On July 19, Mr. Joseph S. Ward, now and for many years past a resident engineer on the Philadelphia and Reading Railroad, wrote to his parents at home: "One man died this morning. I just saw them dump him into his coffin. Another either committed suicide or fell overboard. One died last week. … Mr. Collins is very sick."

Again, he writes on July 24, "I think Mr. Collins is disgusted with the folks at home not sending him any money. His hands are tied. He can establish no credit among the people, nor can he get anything without money and I don't believe there is a cent here. Mr. King is quite sick with fever."

We learn from the same source that on July 27:

> All hands here are sick of the way things are going and are getting very tired of bread and molasses three times a day. Three more men have died since I wrote you last. This makes 17 deaths to date and consequently we feel blue. Of the 700 men who came down I don't suppose more than 200 or 300 are working, the rest being sick. ... We can't get a cent of money even to pay for washing.

The next day, July 28, Mr. Ward writes:

> Another poor man died last night. It is perfectly frightful! The sick men are penned up in a little wooden shed, without ventilation, and lie there sweltering, with their companions dying before their eyes and without any decent nourishment. I don't care how strong a man is, if sick, he cannot get well on such food as salt meat, beans, molasses and bread made of mouldy flour. The men belonging to the engineer corps fare slightly better. We don't allow them to go to the hospital, but take them into our own quarters and buy them whatever delicacies, such as canned milk, etc., can be obtained at the native store. ... We have six or seven sick men at headquarters, who, owing to want of medicine and proper nourishment, are improving very slowly. The worst case is that of young O'Connor.
>
> I tell you it was a good thing you sent that quinine. There is none in the drug store and it is almost indispensable here.

Fortunately, the "young O'Connor," mentioned above, survived, and from his excellent diary, under the same date as that of the letter from which the last quotation was taken, we find the extract which follows.

> Down in the mouth would fitly characterize the feelings of the crowd to-day, but, as it is Sunday and supposed to be a day of rest, we try to snap our fingers at dejection, chills, pills and all other ills of a foreign clime. Our weakness is largely in the legs. I can sit here in a bunch, write and figure away from chill to chill rather than from meal to meal.
>
> Many of us dare not move about, as internal conditions will not permit it. Even when we lie in our hammocks we do not allow them to swing. During the night I notice perspiration increases. This morning my pillow was wet clear through. Even the old bony head perspires a hundred fold more than the rest of the body.

Five days later, Mr. O'Connor adds: "Very low the past few days. Doctor gave me sleeping powders almost every night. Perspiration increases and now becomes a nuisance. The damp clothes, no doubt assist Mr. Chill to get in his work."

Under the date of July 31, Mr. Ward writes, "Ben Huff, who had a contract, died last night," and on the following day, he says in another letter, "Our old cook, Mike Rodgers, died of chills and fever last night."

Mr. Ward's next letter, written on August 2, tells of a narrow escape made by the occupants of the old headquarters building. It is only necessary to say by way of explanation that months before, much timber had been felled and brush cut down in the location referred to. Otherwise, an extensive conflagration in the forest would have been impossible. He says:

> We had an exciting time here yesterday afternoon. The woods behind headquarters caught fire and the flames spread rapidly, burning for a long distance. As the powder house was only 300 yards distant and contained six or eight tons of powder, every one was anxious. Finally men, who were trying to remove the powder, rushed down the hill, dropped their loads and cried "Run for your lives! Run!! Run!!!" Every one in San Antonio ran. The sick were carried about half mile down to the river. Fortunately the magazine did not explode, but, when I went up in the evening to look at it, our escape seemed almost miraculous. The woods around were burned and the powder house was scorched. Only two or three open kegs, which had been removed to some distance from the house, exploded.

On the third, we find Mr. O'Connor indulging in rhapsodies over a meal of ham and eggs sent him by Mrs. Collins, who, like Mrs. Nichols and Mrs. King, will long be remembered for many such acts of kindness to the sick. Mr. O'Connor closes his reference to the subject by saying, "Never while I live will I ridicule a meal of ham and eggs." Under the date of the seventeenth, he further informs us that "To-morrow, if I can brace up, make a bold front and manage to walk erect, I hope to be allowed to start homeward on the steamer *Jurua*. It is now some ten weeks since I returned to headquarters. I am only one of many. No doubt there are a number still encamped in the woods, that are just as profitless to the contractors."

The *Jurua* departed on the eighteenth with Mr. O'Connor and several other members of the engineer corps on board. As usual, Mr. Ward availed himself of the opportunity to send a letter home, from which the following extract is taken.

> I send this by Mr. Delleker, who leaves for home to-day, having arranged with some others, after great difficulty, in getting transportation to Pará. From there they will have to depend upon chance to get home, as there is no money here. In order to reach Pará it was necessary for these fellows to sell watches, sleeve buttons, rifles and whatever other articles they happened to have. ... As soon as possible, send me a few of the most necessary medicines and one bottle of quinine. There is not a grain here and I had a man offer me twenty dollars in gold for one ounce. ... Things are in the same condition, if anything, worse. Grub short, men sick and no money.

Mr. O'Connor's diary concludes as follows:

> I am booked to leave on the *Jurua* to-day. My best wishes for all, who still remain. It is only the weaklings with their energy gone, who are leaving. The mosquitoes, tarantulas and little red devil ants remaining as reminders to stronger men that the fight for blood is still on. I will be in no hurry to return and battle

for life and fortune with machete or transit. Yet I am glad I came. I have no regrets. Time was not wasted. I hope something has been learned. Never again will I voluntarily suffer for wealth without health. We have enjoyed the society of genial companions on the Madeira. Now we part. May we meet again in better health and mayhap in less populous forests, when we will miss the beautiful colors and the many jabberings of macaws, parrots, toucans and monkeys with their white faces and grip tails. Shake boys! May we all meet under pleasanter skies, where health shall be first! May you each bring with you a stronger body and steadier legs than I can boast of to-day! I am feeble in limb, and light in body, having dropped in the forest some thirty and more pounds avoirdupois.

On August 19, Mr. Preston had only been three days in San Antonio, yet under that date, this entry appears in his diary: "Four men have died since I arrived, making a total of 30 since February 18th. Surprised at the few men working. About one-fourth at work and three-fourths sick. Track is laid to station 150." Stations were one hundred feet apart.

In a published letter of Mr. Buchholz to Franklin B. Gowen, this statement appears: "Just before the arrival of the tug *Brazil* with a lighter load of provisions (Aug. 27, 1878), we paid twenty-five dollars a barrel for flour to one of the Brazilian merchants established at San Antonio."

On August 28, Mr. Ward writes:

> The *Brazil* arrived here yesterday afternoon, having in tow two lighters containing a portion of the cargo of

the *Ida M. Eldridge*, which they were obliged to unload at Pará. This arrival was very opportune, as we were about out of grub, entirely out of medicine and had to buy flour from natives at twenty-five dollars per barrel. We now have on hand a supply of superior provisions, consisting of canned goods, etc., which were especially needed, as scurvy was getting very prevalent. ... There is still a great deal of sickness and 34 men have died, but there is now hope of improvement. Five hundred natives of Ceara are on their way up the river to work on grade. Accounts of financial matters, both in England and at home I understand are quite satisfactory.[25] We have received no money. I believe some came up on this boat, but not enough to pay off.

The *Brazil* started back to Pará on the twenty-ninth, and a few hours later, Mr. Hepburn prepared a letter for Mackie, Scott, and Co., giving his impressions regarding affairs at San Antonio. In it, after some statistics showing the actual work done, he pays this well-merited tribute to Mr. Collins:

> Mr. Thomas Collins's personal courage and irrepressible energy certainly deserve the highest encomiums. Contending with difficulties that would have appalled most men, he forced the work ahead with unpaid, poorly fed and disheartened laborers. In constant doubt as to the continuance of his supply of provisions and, on two occasions, only saved from extremities by the opportune arrival of the tugs with their tows, living in

---

[25] This was an incorrect inference, drawn from the improved character of the supplies.

an atmosphere charged with threats and denunciations directed against himself, surrounded and hemmed in by doubts and uncertainties of the gravest character, he, nevertheless, stood firm and immovable as the adamantine rock and prevented disintegration from ending in absolute dissolutions. He has made mistakes perhaps, but no mortal is infallible and he has been tried continually to the extreme limit of human endurance.

Mr. Creighton's diary, under the date of September 7, contains this statement:

> At 9 A.M. saw savages on the path leading to old headquarters. Had to go back by the new headquarters and get home that way. That this was no false alarm is abundantly proved by later occurrences in the same locality.

Mr. Hepburn took with him on the return trip to Pará seven engineers and a dozen or more laborers. On September 8, at 2:30 p.m., Patrick Gorman unexpectedly died on one of the lighters. In that climate, a corpse could not be kept above ground more than a very few hours and arrangements were at once made to have the funeral take place that afternoon. Mr. Hepburn was anxious that the obsequies should be conducted with all due formality, but there was no clergyman on board and no one so preeminent to piety as to make the choice of a proper person to conduct the burial service as an easy matter. After consultation with one or two others, a gentleman, whom we shall designate as Mr. Blank, was chosen to officiate, because of his generally circumspect behavior and the restraint he usually placed on his tongue under trying circumstances, which caused others to express their thoughts in "Language plain and terse, but much unlike a Bible verse."

With some hesitation, Mr. Blank agreed to act, but only on condition that Mr. Hepburn would stand by and, at the words "Dust to dust and ashes to ashes," throw a handful of earth on the coffin. It was 5:00 p.m. and the sun near the horizon when the mournful procession reached a point on shore, where a grave had already been dug. At that hour, the piums and mosquitoes were terrible, and all the time Mr. Blank was reading the service, he and his auditors were subjected to almost unbearable torture. When the words were pronounced that were to give Mr. Hepburn his cue, that gentleman had entirely forgotten his part and was dancing about, now on one leg, now on the other, and all the while vigorously beating both sides of his head with great bunches of leaves held in each hand. There was a momentary pause, as Mr. Blank contemplated the unseemly antics of his coadjutor, and then, bursting with indignation at the unnecessary prolongation of his own sufferings, caused by Mr. Hepburn's failure to perform the simple duty assigned him, he said in a quiet, but aggrieved and petulant tone, *"Hepburn! Why in hell don't you throw on that dirt?"* [Mr. Blank still lives and is an elder in a prominent Eastern church.]

On September 14, a Bolivian Indian who was making a canoe in the woods behind San Antonio was attacked by savages and severely wounded.

The steamer *Canuman* arrived at San Antonio on September 17 with between four and five hundred laborers, who had been recruited in the province of Ceara, on board. They were the poverty-stricken survivors of a smallpox visitation, which had proved extremely malignant and carried many to their graves.

By the regular mail steamer *Javary*, on the twentieth, Mr. Ward writes to the effect that the original force of nearly seven hundred men had been reduced "by death, running away and legitimately leaving for home to about 150 or 200. ... The general condition of things about the same. ... Not much sickness but a great deal of discontent among the men because there is no money."

To illustrate the methods by which some of the laborers at San Antonio reached their homes, it is proper at this point to introduce, in its chronological order, an extract from a published article, based upon an interview with Professor S. C. Hartranft of the Northern Normal and Industrial School at Aberdeen, South Dakota, who, when a lad of eighteen, was one of a party of axmen employed on the work at San Antonio.

> His party was composed almost entirely of young fellows from Northumberland County, Pennsylvania, ... under command of a man named Huff. ... Compelled to work in the noxious vapors of such a climate, to sleep on the wet ground teeming with malarial dampness, to eat, after a day's work, the insufficient and unwholesome food, could have but one end, sickness and death. ... Despair and desperation began to grow upon the party and many of them stole off to find their way back to

their homes. ... Hartranft and a party of ten others resolved to leave. They had worked their six months, fulfilled their part of the contract and, besides, had obtained permission to go. They purchased a boat and equipped it with several barrels of hard-tack and some salt meat. They had no money after buying the outfit, as the contractors had not paid their men a cent, but promised payment by their agent in Pará. On the night of October 2d they left San Antonio and for nineteen days rowed down the Madeira and the Amazon rivers, a distance of 691 miles, to Serpa, a small town of about 700 inhabitants and a custom house. During that time the provisions had given out and there was discord as to the party's future course. The men's nerves were unstrung by hard exertion and their weak condition at the time of starting.

The nights spent in the boat on the river were uncanny, the darkness being filled with the chattering of monkeys, the roaring of tigers and other terrifying sounds. One of the men named Carpenter, from Ohio, because drunk and insubordinate, was left on the bank of the Amazon. That was the last known of him. At Serpa the party hailed a steamer, sold their boat and some trinkets and had just eight dollars apiece, after paying eleven dollars each for passage to Pará. They were treated as slaves, fed a slave's fare and slept with the cattle on deck. When they arrived at Pará the agent told them sadly that he had received no money for them. They were in desperate straits with no money and far from friends; so they applied to the American consul for aid. That dignitary told them it was not the business of the Government to look after all the tramps roaming about the country, but that if they could find no means of sustenance he would see what he could do. At last he found them a place to sleep at an Englishman's hotel.

They were boarded free of charge at an eating place, kept by a man, who had previously deserted from the railroad. At the end of a week they were kicked out by the Englishman, because they could not pay for their lodging. Then they went to a native hotel and there remained until November 15th, when they boarded a two-masted schooner for New York. One of the persons on board was Horace Aunxt, a boy neighbor and friend of Mr. Hartranft, whose home was in Milton, Pennsylvania.

The boat, delayed by tropical calms and storms off the coast of the United States, did not reach New York until the day before Christmas. The captain, Jesse R. Cavalier, became disgruntled at the men, because he found the orders they had given him were worthless. He went ashore, leaving them aboard ship for two whole days without fire and food. Still dressed in tropical clothing, they nearly froze to death and their Christmas dinner that year consisted of some cold boiled potatoes.

Finally they succeeded in making a landing and their appearance in New York City was an odd one. They had a letter to a business man in Maiden Lane. He helped them as much as he could, but Mr. Hartranft says to-day they would have frozen or starved to death, had it not been for the saloons. They sold some pieces of tropical wood to pay their fare to Philadelphia, where they arrived December 27th. They met Mr. Philip Collins at his office, but he was bankrupt and could not pay them for their work. The first really good meal the men had, after leaving home in February, they enjoyed with the servants of a Philadelphia hotel. The youngsters ate well into the night and, next day, were furnished by Mr. Collins with transportation to their homes.

October 7, 1878, was marked by two events, an arrival and a departure, both of which made a profound impression upon the Americans at San Antonio. The first was a subject for almost universal rejoicing, and the second was the cause of more general and more sincere sorrow than almost any other of the many misfortunes which this narrative records.

On that date, Mrs. King presented her husband with an infant daughter, afterward known as Miss Maria Juanita King. She was their first child, and Mr. Ward, in one of his letters, truthfully says, "The boys are quite jubilant over the new arrival of their Amazon Princess and take a lively interest in her."

It requires long months of residence in a camp where a baby's cry is never heard for anyone to appreciate the flutter of excitement among the male sex caused by Miss King's advent. At home, the subject of babies would have probably had no interest whatever for most of our men, but at San Antonio, it was different. Certainly, no congress of mothers ever cackled more over their own babies than did the men at headquarters, most of them young bachelors, over Miss King.

The steamer *Canuman*, which left San Antonio on September 18, had taken as passengers, eight members of the engineer corps and among them Mr. Rodman McIlvaine of Philadelphia. Mr. McIlvaine was unwell when he left San Antonio and continued to grow weaker on the voyage.

Above Santa Rosa, the *Canuman* grounded on some rocks and lay there for two days until the mail steamer *Javary* came along. As Mr. McIlvaine was in urgent need of medical attention, Mr. Wetherill had him transferred to that boat and, with him and others, arrived at Manáos on October 7. The best physician obtainable was at once summoned, but it was too late and McIlvaine passed away before medicines that had been sent for arrived. He was buried in the Catholic cemetery at Manáos, and a monument, erected there by his comrades, attests to this day the high esteem in which he was held by all who knew him. Mr. McIlvaine and the writer had jointly occupied the same stateroom on the *Mercedita*. At that time, his good habits, excellent general health, and splendid physical development seemed to be guarantees of unusual capacity to endure the privations and hardships of camp life. He was almost continuously in the field from February 19 until, with Mr. Wetherill's corps, he reached the Jaci Paraná on August 8. His field service had, therefore, been exceptionally long and arduous.

It is, however, a striking fact, of which we had many proofs, that malarial fevers are unusually malignant in cases where men have previously enjoyed robust health, while those threatened with consumption or affected with ailments that would apparently unfit them for efficient service, often exhibit astonishing powers of endurance.

In a letter dated October 15, Mr. Ward writes from San Antonio that, on account of the arrival of the laborers from Ceara, "work goes on a little more briskly than heretofore. ... But a very small percentage of the Americans are working. ... A great many have gone home and many are now leaving in small boats they build for themselves. Only sixteen engineers now remain." But there was no abandonment of the enterprise or even thought of it at that time.

A contract still exists, bearing the date of October 17, 1878, in which the *commandante* at Praia de Tamandua agrees to furnish Mr. Collins with a thousand turtles, which were soon afterward delivered alive. They were of immense size, some of them measuring three or four feet from head to tail. Eaten in moderation, they were a real treat to men who had for months subsisted upon fat pork and salt beef, but the trouble was that, while the turtles were obtainable, we had a surfeit of them. The writer, then in very bad health, had arranged with Sr. Brigido, who kept the principal store in the place, to furnish him with table board at a price equivalent to twenty-two dollars per month in American money. Sr. Brigido rated the quality of his table fare entirely by the number of courses served, and it was no uncommon thing to have turtle or turtle's eggs, cooked in different ways, served five or six times at each meal. The turtles were only obtainable in large numbers during the extreme low water of September and October, when they crawl upon the sandbars by the hundred at night to deposit their eggs in the sand.

A canoe filled with Indians noiselessly floated down to the hatching ground after dark, and it was then only a question of how many turtles could be turned on their backs before they escaped to the water. The eggs were taken out of the sand, some of them half hatched by the heat of the sun, broken by treading on them, and the mass allowed to settle. The oily part soon rose to the top and this, skimmed off, constituted what the natives called turtle butter. We were inclined to accept, without verification, Keller's opinion of this delicacy. He says it has "an abominable flavor, thoroughly disgusting to a civilized Christian's palate." It is only just to Sr. Brigido, however, to state here that the eggs served on his table were unlaid ones, taken from the turtles after they were killed.

Only Mr. Collins and Mr. Nichols had at this time any clear idea of the real situation in England. The others were mainly dependent for information upon rumors carried by officers of mail steamers from Pará. These rumors had been very conflicting and we were constantly changing from a state of confidence in the issue to one of utter hopelessness, or vice versa. Early in November, however, it became evident that it was useless to expect the contractors to pay their employees until questions raised in England were settled by litigation, which might consume months or even years of time.

The realization of the fact that they were sacrificing health and life without either pecuniary gain or reasonable hope of being able to complete the railway led many to arrange for departure at the first opportunity.

About the same time, Mr. Collins heard that several Brazilian engineers, who had been watching our progress for the Brazilian government, were preparing a report and were inclined to recommend that Brazil should come to the relief of the contractors pending a legal decision on the case at issue in the English courts. Mr. Collins at once decided to go to Rio de Janeiro in order to see that the facts were correctly stated to the emperor but delayed his departure until the arrival of his brother, Mr. Peter Collins, then known to be on the way from Philadelphia.

The steam launch, *Carlo Lima*, which brought Mr. Peter Collins to San Antonio on November 12, the next day, carried away Mr. and Mrs. Thomas Collins, bound for Pará and Rio de Janeiro.

That same day, three Bolivian Indians, while engaged in chopping wood near the powder-house, in San Antonio were attacked by savages; one of them died soon after with seven wounds, but the other two, though severely wounded, recovered.

By the steamer *Anajas* on November 20, Messrs. Preston and Clark, of the engineer corps, and several laborers departed.

On the twenty-seventh, Mr. Ward writes:

> Things look as if we might have trouble with our Ceara laborers. I fear they are a bad lot. Mr. Peter Collins (then only fifteen days in San Antonio) has already had a dose of the fever, but is recovering. ... All hands (at headquarters) are down sick except the two draftsmen.

The next day, the steamer *Theotonio*, with two empty lighters in tow, sailed for Pará, where she arrived on December 13. Dr. Whittaker and twelve engineers were on the steamer and 199 laborers on the lighters. Five men died on the way, and few of the others reached that place in a fit condition to take care of themselves. "Old Mike," though no longer employed by the contractors, assumed full charge of all invalid laborers during the trip, made the five coffins, and read the burial service at each of the five funerals. Even at Pará, when nearly all who were able to do so went ashore, he would not desert the large number of invalids who remained on board the lighters, though he knew their supply of provisions would be exhausted in two days. The writer knows not whether the once sturdy Irishman still lives or whether he has gone to join the comrades by whom he stood so faithfully in sickness and in death. He has long since forgotten his full name, if indeed he ever heard it, but no name, however illustrious, could increase the profound respect he has always maintained for the rugged simplicity of character and sterling virtues of "Old Mike."

Before Dr. Whittaker sailed for New York, he succeeded in getting some of the worst invalids admitted to the municipal hospital in Pará, which he himself said was "a worse place than hell." Others called the attention of our consul to the nearly two hundred destitute American laborers, most of them sick, ignorant of the language of the place, and without hope of obtaining more than 7 percent of the money due them, as shown by orders they held on the agent of P. and T. Collins at Pará.

After the departure of the *Theotonio*, on November 28, the working force on the railway was reduced to 11 engineers, about 150 American laborers, and about 325 natives of Ceara. A letter of Mr. Ward's, dated December 11, says:

> We have hardly any men working. ... Have not been able to prevail on many of our Ceara laborers to return to work. ... Mr. King yesterday had only six men on grade. ... The chief engineer has to be almost fighting all the time to have anything done.

Mr. O. F. Nichols, the resident engineer for the Madeira and Mamoré Railway Company, was then receiving confidential communications from Colonel Church in London. On two occasions, these letters had been opened in transit through Brazil. In one of them, he had been advised that he might be called to London to testify in the pending litigation. He determined, therefore, to decrease the distance between himself and the cable office by going at once to Pará.

Leaving his assistant, Mr. Franklin A. Snow, to represent him at San Antonio, he sailed for that city on December 17, where he arrived about two weeks later.

Most of the 212 passengers who came to Pará by the steamer *Theotonio* on December 13 were still there when Resident Engineer Nichols arrived. The total number of destitute Americans in the city then fell little, if any, short of three hundred men, and Mr. Nichols thus describes the situation, as he found it.

> These distressed and homesick fellows, waiting patiently for transportation, were natives or residents of the United States and their condition in Pará was truly pitiable. Without adequate clothing, with no money and no shelter they literally became beggars on the streets. Two rooms were secured for them in the second story of a building, where they might have lodging at

night, and subscriptions were taken up by the American Consul and others, among the citizens of the town, to provide them with food for the two or three weeks that must elapse before the American steamer would arrive to take them home. Enough money was raised to provide them with one meal a day for this period, but many of them insisted on having this fund expended at the rate of three meals a day, trusting to luck and begging to furnish them with food after the fund was exhausted.

The parties in charge of the subscription, however, held firmly to the original intent and furnished them only one meal a day for upwards of two weeks. More than half of these fellows would beg for money, and some of them used it for the purchase of liquor, of which they could buy for a couple of cents sufficient to keep an able-bodied man drunk for half a day.[26]

They got sick, many of them went into the hospital and some died there.

One of the saddest incidents in the life of the writer was his experience with poor Rulon Gray of Chester, Pennsylvania, a big, strong fellow, about twenty years of age, who had gone down to Brazil on the *City of Richmond* and was stranded in Pará. He took typhus fever in malignant form, was sent to the hospital and, on the last visit paid him, was yellow with the disease and the picture of death itself. The steward said he could not live, and the writer promised to

---

[26] The best quality of *cachaca* sold for twenty-five cents a gallon or its equivalent in Brazilian money.

see him again next day. On calling they said he was dead and buried. This could not be credited and later investigation showed that his body had merely been placed in the morgue for burial in the potter's field. Remonstrance was of little use unless accompanied by some offer to provide for the interment, but this was quickly made and Gray's body, which had merely been sewed in a white sheet, was transferred to a neat coffin, a burial permit obtained from the authorities for interment in the English section of the cemetery, and there, attended by one coach, containing two persons, poor Gray was buried, and the key to the coffin sent to his parents in Chester.

The steamer finally came and all the poor fellows in Pará were taken on board for New York, where they arrived, after a voyage of about ten days, and disembarked in the month of January, a sorry looking crowd, most of them attired in the lightest of summer clothing and many wearing straw hats and linen dusters in midwinter.

Returning to San Antonio, we find that, soon after the departure of Mr. Nichols, Mr. G. W. Creighton, with two other engineers and about twenty Indians, was sent to collect camp equipage, left at various places on the river. During this trip, he had occasion to stop all night at La Concepcion and thus describes his visit.

As we approached our ears were greeted with the sound of music and our boatmen hastened toward the shore, anticipating like ourselves a night of pleasure. Instead of that, however, our host (Don Pastor Oyola) met us at the landing place with the announcement that one of his children had died and that his retainers were mourning. With his usual hospitality, we were, nevertheless, conducted to his home, our hammocks hung in comfortable places and we were then led to the large open central room on the second floor, where, amid surroundings almost barbaric in character, lay the corpse of his child. The coffin rested upon a table at one side of the room. The little body was richly clothed and its hands were clasped around the handle of a double fan made of colored paper. The two fans were arranged one above the other; so that the upper one extended a foot or more above the corpse. The coffin had been constructed of rough boards hewn from logs and was completely covered with white paper, upon which strips of colored paper, about half an inch wide, had been pasted; so as to cross and re-cross diagonally and form squares, of about two inches on a side, in the center of which were pasted little crosses of colored paper. The whole was neatly finished and geometrically regular, but the effect was oddly incongruous, though strangely pathetic. Upon the opposite side of the room were seated the musicians, with drum, flute, cornet and violin. Upon the floor, seated Turkish fashion, were possibly two dozen Indian women robed in simple white garments. With bowed heads, they wailed and wept and swayed their bodies to and fro with every cadence of the music. When darkness fell about the room, the flickering light of the candles, the odor of incense, the barbaric drone and din of the musicians and the wild cries of the mourners all combined to fill with awe a stranger unaccustomed to such scenes. Occasionally a long pent-up cry from the interior of the house betokened the presence of the mother, grieving, as only mothers can grieve, over the loss of their little ones. Hour after hour through the whole night this performance continued without cessation. As the musicians tired or the mourners swooned, others took their places and the moans, cries and discordant music never ceased until daylight. Then, or soon after, the weeping parents bade the little one farewell and the coffin was closed. Four Indians clothes in white carried it from the house, preceded by the musicians and followed in double file by the entire population of La Concepcion. The burial was soon over and we returned to the house for breakfast after which we left for San Antonio.

Under the date of January 15, 1879, Mr. Ward again writes as follows:

Things at San Antonio cannot be much worse. ... A great number have been sick and two or three have died. The few Ceara laborers, we have left, are dying off fast. ... Such squalor and filth as exists among them, you can't imagine. There is no work of any kind

> going on, though we can manage in the office to find employment. Bolivian and Brazilian settlers complain of the way in which the contractors treat them. Men, who wish to go home, cannot get money to leave and, to make things worse, we received news that nearly all the men, some 200 in number, who went down in November, are still lying at Pará. ... Men here are selling their clothes, watches, etc., to get away and things, taken all in all, are discouraging.

Two days later, Mr. Ward says:

> I am worried almost to death. ... Our letters from all we hear, are utterly neglected in Pará and are given to any one who asks for them. There are so many blackguards there that we have no certainty of getting any. ... There are also, I understand, boxes in the Custom House at Pará, which I cannot get. There are said to be some forty boxes there for private individuals. ... It is reported that small-pox came up on the boat, but, thank God, I am well and strong. ... All I am afraid of is this small-pox. There is a great deal at Pará, as well as at Manáos and they had, I believe, two cases on the last boat, but they cannot land.

Mr. Vasa E. Stolbrand, one of the most efficient of the younger men on the engineer corps, unexpectedly, even to himself, took his departure on January 19, in consequence of the news that he had been appointed a cadet to the United States Military Academy at West Point. Today, after distinguishing himself for many years in the public service, his name appears among the colonels on the retired list of the regular army.

The medical report of Dr. E. R. Heath, covering the last week of January 1879, is now before the writer. At that date, there could hardly have been more than 120 Americans at San Antonio, yet this paper shows that sixty-six persons had been either in the hospital or under treatment during the week. There we find many familiar names, some of them frequently mentioned in previous pages of this narrative, such as Peter Collins, C. S. d'Invilliers, W. L Symmes, C. L. Moore, Charles F. King, J. Dougherty, Fred Lorenz, W. S. Nevins, Charles La Mott, C. J. King, and "Baby King." As Miss King was then only three and a half months old, it is a gratifying evidence of early precocity to find the complaint for which she was treated stated as "teething." Taking into consideration the large number of persons who were usually their own physicians, this report goes far to prove that few persons in San Antonio enjoyed good health during the last days of January 1879.

While the condition of affairs at San Antonio was daily growing worse and circumstances beyond control seemed to render further progress on the work impossible, there was one man who never gave up the fight, never lost faith in the ultimate success of the enterprise, and never relaxed his efforts to achieve victory over the almost insurmountable obstacles that opposed his progress, until he could no longer obtain healthy men to replace the invalids sent to headquarters from his camp and until his physical strength entirely failed to respond to the demands made upon it by his own dauntless courage and inflexible will.

That man was "Joe" Byers. With a few Bolivian Indians and such Americans as he could obtain, he had, since September, established and marked upon the ground the final railway location for a distance of nearly twenty-seven miles. Leaving his camp on the Jaci Paraná during the latter part of January, he came to San Antonio for the purpose of obtaining provisions and exchanging some invalid assistants who accompanied him for others in better physical condition.

Returning to camp after a round-trip of 102 miles by canoe, without the desired supplies and with Mr. Ward as his only recruit, he found five of his white men unable to leave their hammocks and his provisions limited to canned beef, coffee, and limes. Assisted by Mr. Ward and the Indians, he moved camp some miles northward and then endeavored to continue his work. As the sick men failed to improve under the medical skill and attention Mr. Byers was able to give them, they were made as comfortable as anyone could be in such a country with insufficient food and only leaky tents to protect them from the heavy rainfall, then of daily occurrence. Rising in the morning at daylight, he would himself cook breakfast for all, and then, leaving a bucket of limeade where the sick could reach it, he would go out with Ward and the Indians, run transit line one day, and on the next, run over the same ground with the level. This was kept up for some time, but at last, he was himself prostrated by a severe attack of chills and fever, which did not yield readily to the usual treatment. His assistants were in little, if any, better condition than he was, and it soon became evident that even the indomitable energy, the inflexible determination, and the iron nerves of Joe Byers were unequal to the task he had undertaken to perform. Even then, it was with sorrow that he yielded and abandoned his work. It was no easy matter to get his men to the river. Some of them had to be carried, but all finally reached San Antonio on February 16, 1879.

The next day, Mr. d'Invilliers, Mr. King with his family, and Mr. Ward embarked on a river steamer for Pará. Mr. d'Invilliers had obtained a leave of absence some time before, and as his testimony was desired in the English courts, his departure was no surprise. It nevertheless marked, for the time being at least, the complete collapse of the undertaking.

Messrs. Collins, Byers, Creighton, Patterson, Brisbin, Lorenz, and others remained at San Antonio for several months afterward, some because it was necessary to care for valuable property, others because they hoped the legal obstructions interposed would not long prevent prosecution of the work, but the great majority because they had no means of getting away.

On May 3, Mr. Peter Collins and a man named George Gray were shot by savages near the old headquarters building in San Antonio. Both were seriously wounded. Mr. Collins had been struck by two arrows, one of which penetrated his lungs and caused hemorrhages, which for a long time made his recovery doubtful.

Finally, on August 19, the entire American colony at San Antonio, then numbering only about sixty men, was sent for by the contractors in Philadelphia. With the exception of Mr. Charles B. Brisbin of the engineer corps, who remained for some years in Pará, all safely reached New York on September 26, 1879.

XXVIII. Mr. Mackie's Experience at Pará

> *The men who have really done things that last have been in a large sense men of faith; men who did not let the visible things of the present get out of proportion to the possibilities of the unseen future. To them and to them alone was it given to endure to the end.*
>
> —President Hadley

Though different firms were entrusted with the execution of the railway and navigation projects, it must not be forgotten that they were essential and inseparable parts of one great enterprise. It is fitting, therefore, that we should conclude this store of information about the American operations.

Robert H. Hepburn

Mackie, writing from memory after twenty-five years, slightly mixed in some details.

My records show that Lawford, acting as my secretary, and I—with the officers and crews of the tugs—were the only ones of our outfit in Pará when starting for San Antonio with the *Juno* and schooner *Anthony*.

The others of my corps came out on the same steamer with Davis, while we were on this trip.

"Jack" Pennington came up on the *Javary* for me while I was on the USS *Enterprise*, anchored off Araras, where I had left the *Juno* and *Anthony* returning to Pará to go on board to compare Madeira River work and reported the serious conditions in Pará. I went back to San Antonio, picked up Lawford there, and the three of us returned to Pará—all on the same steamer.

The "fun that began"—to which Mackie alludes and which started after the arrival of Davis—and the "hell that broke loose" with the need to straighten things out, continued to the end.

After "extricating" the *Brazil* from the "shotted-towline" custody of the revenue cutter, we took in tow the two lighters and started for "San Anton"—Keasby and Corham accompanied me—where and when the completed "surveying, sounding and searching" (later plotted by Keasby) was done.

Mackie arrived while we were up the rivers on this trip.

It was the big whistle of the *Brazil* that blared out our return.

The deplorable and distracting conditions succeeding the arrival in Pará of squads of sick and destitute men from the railroad are indescribable; they were awful.

Remembering our life in Brazil is best done by relating the experience of Mr. C. P. Mackie, while stationed at Pará as executive officer of Mackie, Scott, and Co., Ltd. In addition, this we are fortunately able to do by simply resuming the statement of Mr. Mackie's personal recollections at the point where it was discontinued at the end of the thirteenth chapter.

When Mr. Hepburn left Pará, June 7, 1878, on his first trip to San Antonio, he had the tug *Juno* and a tow consisting of the three-masted schooner *D. M. Anthony* laden with seven hundred tons of railroad iron, coal, and provisions and a two-hundred-ton iron lighter full of coal. The schooner drew fourteen feet of water, and everyone in Pará who knew anything of river navigation predicted the complete failure of the *Juno* to take such a tow to San Antonio at that season of the year. These predictions were based upon the supposed inability of such a diminutive tow boat to stem the currents of the Madeira, which in many places reached their maximum strength during the season of low water, and upon the belief, entertained by the best river pilots, that no craft that drew more than eight feet of water could navigate the Madeira at that time.

No one more incessantly condemned the foolishness of Mr. Hepburn's attempt than Mr. Fred Pond, the head of the only American mercantile establishment in Pará, then temporarily acting in place of Captain Lima as financial agent for P. and T. Collins. This discouraging testimony so wrought upon the fears of Captain McLane of the *Anthony*, that, as Mr. Hepburn says:

> He used to come to me in all kinds of moods, varying from bluster to appeal, urging me to permit him to discharge his cargo at once and not compel him to risk losing his vessel in a rash attempt to take her to San Antonio, but I emphatically refused to comply with

his wishes and peremptorily ordered him to be ready to start at an hour's notice.

The opportune arrival of the *Juno* and her tow at San Antonio on June 27 has already been mentioned.

It was probably while Mr. Hepburn was absent from Pará on this trip to San Antonio that the senior member of Mackie, Scott, and Co., Ltd., reached that place. Without further prelude, we now submit a second extract from Mr. Mackie's interesting paper.

> About the time of my arrival the fun began and it was variegated and intense. Hepburn had gone up the river with the *Juno*, towing a schooner and a barge—a task held to be utterly crazy in the opinion of every experienced river man in Pará—taking with him most of the members of our expeditionary force. His and their courage, skill and sound sense established the feasibility of such methods of transportation over the 1683 miles, between the Atlantic and San Antonio. If nothing else resulted, the present rubber trade of the United States is a tribute to the daring and good judgment of that bunch of young Americans. A comparison of any chart of the Amazon and Madeira rivers, made before 1878, with those subsequently published, by either the U.S. Hydrographic Office or the British Admiralty, will be sufficient to convince any one that those youngsters did things.
>
> West, as general agent and manager, had been left in charge at Pará to attend to our relations with the Government and with the general agent and attorney

of P. & T. Collins, as well as to oversee the putting together of the *Santa Maria*, the *Beni* and the *Mamoré* in the navy yard there, with facilities granted by the imperial Government of Brazil.

The *Santa Maria* was a light draught exploring launch of high power and had a tendency to swamp at full speed; the *Beni* was a stout river tugboat, intended for towing purposes above the falls of the Madeira; the *Mamoré* was a light draught trading steamer with powerful engines intended to penetrate all the upper waterways of Bolivia and collect their produce for shipment to San Antonio and thence, *via* Pará, to Philadelphia.

The general agent and attorney of the Messrs. Collins at Pará was Captain Lima, for many years a commander in the service of the Pacific Mail Steamship Company. A Brazilian by birth and a gentleman by instinct, Captain Lima was as full of the bristles of personality as a porcupine is of quills. The inspector of customs at Pará was a Brazilian gentleman, who had been educated in Salem, Massachusetts, spoke English perfectly, was the essence of all courtesy, but could never get it out of his head that Mackie, Scott & Co., Ltd. and the Messrs. Collins were filling the empire of Brazil with contraband goods, introduced under the authority of an assignment made by Colonel Church to us of the privileges granted in his concessions, that such articles were to pay no duty, would make us very rich and would do him no good.

As to the three boats and their erection, any one who has worked in South America with native workmen under Government control, within ninety miles of the equator, and just about a thousand from a machine shop, can imagine what poor West had to go through in the performance of his duties. But, while Hepburn and his fellow pioneers were taking the *Juno* and her unprecedented tow up the rivers, surveying and sounding them and fixing his astronomical points as he advanced, West struggled with the entirely unique experience confronting him at Pará, and, in the words of the old New England tombstone "done noble." Both men had, in good plain English, the very Hades of a job and they and their respective associates virtually worked for the love of success, certainly not for pecuniary profit.

You will remember that after one enters the one-hundred-and-fifty-mile mouth of the Amazon and passes up the Pará River outlet of that great stream, by and by the towers of the cathedral at Pará come into view. Not long after I saw a handsome tug of the Delaware River pattern, bearing the proud name of *Paul Mackie* painted on her stern. This was hailed by my companion as a good omen, but I wondered what in blazes such enthusiasm meant; because I was only a small frog in an enormously large puddle. It turned out to be the tender of the Messrs. Collins, thus named by their attorney and general agent, Captain Lima, from a quite mistaken idea of the writer's importance in the enormous undertaking, in which we were all involved. Two days afterwards West called my attention to the same tug lying in front of our hotel at

Pará and handed me a pair of binoculars, with which I descried that her name had been changed to that of the *C. de Lima*. In the meanwhile things had become different. Such is fame!

Our contract with the Messrs. Collins was extremely specific and clear. We were to do certain work between Pará and San Antonio and be paid a certain price for doing it. The trip, on which Hepburn with the *Juno* was then engaged, meant, if I remember correctly about $100,000 to our firm and Captain Lima got restive over the incident. Each succeeding trip meant about as much; for the contract had been intended, by ourselves at least, to return all expenditures and yield a handsome profit in the first year of its operation. On merely equitable grounds it should have done so, in consideration of the enormous risks, personal danger and unlimited labor involved in the very initiation of such an enterprise. But for some reason Captain Lima, who was very close both to provincial and imperial Government and had the ear of all the under-officials, felt aggrieved or hurt by something or other, and, from that time until the whole enterprise broke down and we were notified to return to the United States, we were up against difficulties and obstacles, in comparison with which the mysteries of navigating the Amazon and Madeira and the physical obstructions presented by the latter's 240 miles of falls and rapids, were not worth mentioning.

West and I, with the others, labored to get our three river boats into the water, and still more to create an atmosphere of co-operation between ourselves and those persons who represented the Empire of Brazil, the Province of Pará and the Messrs. Collins. It was a rather complicated situation and doubtless we made many mistakes, but, to cut a long story short, the time came when West and I, sitting in the hotel facing the river at Pará heard one night the private signal, which I had established for ourselves in Brazil and Bolivia. It was the old crossing signal, at one time used exclusively by the Pennsylvania Railroad, two long whistles and two short ones. It meant to us that the *Juno* was back safely from her long and perilous passage and that we should know how things were going on at the big camp, which the Messrs. Collins had established more than 1500 miles above Pará.[27] Then hell broke loose; way up there at San Antonio, with 700 men pitched on the banks of a tropical river, pretty much as the steel rails we had towed there were thrown, disease was rife and deaths were reported to be alarmingly frequent. Many of the employees had appealed to those on board the *Juno* for transportation and a number were actually conveyed to Pará on the empty schooner. Some of those, who crawled ashore, were so weak that they fell in their tracks, and in twenty-four hours the word, American, meant beggar in Pará.

We had two schooners loaded with provisions and railroad material lying in the harbor and for weeks had been vainly trying to get permission from the agent of

---

[27] The *Juno* left San Antonio July 6 with the schooner *D. M. Anthony* in tow. Mr. Hepburn went down the river about four hundred miles with the *Juno* and then returned to San Antonio on the *Javary*, left that place on the same boat July 18, and reached Pará on August 2, 1878 (NBC).

the Messrs. Collins and the authorities of the provincial custom house to send their cargoes to San Antonio with the *Brazil*. Now we tried to get permission to send the railway supplies by the *Juno* and, on account of her greater power, to send the *Brazil* ahead laden with provisions, luxuries and medicines for the stricken camp. We met with a point blank refusal and even our, perhaps unwarranted, offer to send the *Brazil*—then proven to be the speediest boat on the Amazon—at our own expense, with at least remedies and necessities, was refused. As we talked the matter over with the agent of the contractors and with the officials at the Custom House, sitting in the comfortable shade near their offices, we could see sick and destitute Americans, who in various ways had reached Pará, for want of other shelter from the burning sun, seeking the shade of the walls and warehouses along the river front, where some of the more debilitated seemed likely to die. It was none of our business, but we gathered the invalids into the hospital and, at our expense or at the expense of some of us, obtained passage to the United States for many on outgoing schooners.

There can be no question as to the deep interest taken by the Brazilian Government in the success of our enterprise. Neither can there be any doubt that the Messrs. Collins had the welfare of their employees constantly at heart, but the agents and representatives of both parties at Pará were constantly at swords' points with us over questions of no real importance, such as boiler inspection, carrying custom house inspectors on our boats and other matters, which seemed, to men, rendered impatient by the urgent requirements of the service, mere red tape inventions designed to cause delay. This state of affairs almost resulted in open conflict between the subordinates of the three parties concerned. We could move neither the representatives of the Government nor of the contractors and all the time reports that hundreds of our countrymen, good, bad and indifferent were rotting up there in San Antonio very nearly drove us mad.

Finally we decided to take things into our own hands and, although the Customs authorities put a revenue cutter alongside the *Brazil* with shotted guns to prevent her departure, she got off all right on her second trip, went up the river with two lighters in tow, arrived at San Antonio on the 27th of August and contributed largely to the relief of the afflicted camp. In this particular episode the man who stood out prominently was Dr. Jack Pennington. There is little doubt in my mind that his terrible experiences, with disease, squalor, awful heat and frightful airs during the many weeks he was attending to his devoted and unselfish labors among this herd of diseased castaways, right under the equator, in one of the worst climates in the world, sowed the seeds which led to his most lamented death years afterwards in Colorado. Peace to his ashes; for he was a white man!

*

Of a different sort was Lieutenant Spalding's adventure at the mouth of the Amazon. When the *Enterprise* returned to Pará, Commander Selfridge deemed it necessary to fix astronomically the exact position of the little fixed *farol* or light on the north shore of Marajo Bay, about 50 miles above Pará *via* that nasty waste of lumpy waters.

The corvette's launch was dispatched there with three days' *chow* and, when the sixth day came and no returned launch, all hands were perturbed. You may remember that even the largest river steamer would refuse to traverse the bay when the wind blew strongly against the down-coming flood. Well, one night, about 2 A.M., while West, Lawford, myself and others of the M.S. & Co., Ltd. party were trying to sleep in our hammocks on shore, we were aroused by a terrific pounding on our doors, and a fog-horn voice chanting rhythmically "Bullies ahoy, there!" It was Lieutenant Spalding, navigator of the *Enterprise*, with a detail of three shell-backs, one a grizzled quartermaster with "Nantucket" written large across his pickled face, the others able seamen, whose bronzed complexions, husky voices, horny hands and time-worn visages suggested memories of bygone days, when bloodless victories were not of common occurrence in our naval history. Spalding's errand was urgent and his manner imperative. Commander Selfridge's compliments, and he desired to charter the *Santa Maria* to go instantly in quest of his missing launch. Money was no object; he would give us a draft on the Navy Department for any price we named—any price. Anticipating our acceptance, he had sent along the detail necessary to man our boat, as every hour might make the difference between life and death.

Rather rough on us, wasn't it? Even Spalding's eyes twinkled a bit as he mentioned the draft. But as you know by past experience, it is hard to be heroic, standing in your pajamas with a lighted candle in your hands, which the bats are trying to extinguish, while you rub one bare foot against the other to keep the cockroaches from crawling up your legs. Anyway, we soon satisfied Spalding that we were not in the life-saving business for profit, but were very willing to lend Uncle Sam the *Santa Maria* for any such laudable purpose, provided he would take along Lawford, our big and wise Canadian engineer, to run her and myself to represent the firm and act as handy man.

Within an hour we were under way, having picked up at the wharf a supply of provisions, which Spalding had brought from the *Enterprise* in his pull-away boat. We went to the north of the islands to avoid the morning sea, but even so did not reach the north shore of the bay until the following night. Of course, the charts which Spalding carried were of no more use than so many strips of wall paper, and, after blundering into half a dozen inlets and creeks, we finally struck the spider-legged shack, which did duty as light house for the illumination of the world's commerce at the mouth of its greatest river. There we learned that the *Enterprise's* launch had arrived several days before and had departed, after taking the desired observations. A tremendous sea was running in the bay as far as the horizon, and, after making a number of futile efforts to get across, we were thoroughly weather-bound for twenty-four hours.

The *Santa Maria* had only about ten inches of free board and dragged down the stern to such an extent that one comber out of every three would tumble inboard and wet us to the knees, to say nothing of Lawford's fires. Finally, as our provisions were down to two rations and Spalding was intensely anxious to get back to his ship, we decided to take a "bee-line" course across the trend of the seas, attempt to reach the mouth of the Tocantins River and thus work our way by the inside canoe passages back to Pará.

It took us fourteen hours to make the fifty miles and a good many more than fourteen times "Old Nantucket" shook his head as the launch filled with water and wallowed in the trough of the sea. At daybreak one morning we made the mouth of the Tocantins River and ducked blindly into the first lee passage that showed between the hundreds of islands forming the delta of that river. We were wet, hungry, completely exhausted by loss of sleep and continuous bailing. We had no desire to contend against the current of the river any longer than was necessary. Our wood all gone and down to eating monkeys, we were closely scrutinizing both banks of the narrow waterway for a place to land, cut some wood and shoot something or other to eat, when, at a turn of the little canal, we came upon a large and permanent rubber camp, bearing evidence of sustained prosperity and quite a little population.

It did not take long to reach the landing place, where, attracted by our loudly sounding whistle, the proprietor of the establishment, a corpulent gray headed Portuguese, dressed in blue duck, with the usual multicolored assemblage of Indian and Negro men, women and children, was waiting to receive us. On hearing of our needs, the proprietor, with spontaneous hospitality, offered to have his people supply the launch with fuel while we went to the residence and had a good breakfast.

The latter was bountiful and more than acceptable. While the meal was being prepared the old proprietor plied us with questions as to who we were and what we were doing in those remote regions. On being told that we were Americans, he asked whether that was the American flag on the launch. Being assured that it was, he insisted on walking down to her again and carefully counting the stripes and stars, saying that he had never seen the American ensign before, but considered that *Abrão Leencohn* was the greatest man the world had ever seen.

Throughout the meal our host seemed preoccupied and, at its close, he gave some orders, which resulted in the mustering of fifty or sixty of his laborers, the majority of whom were Negroes or mulattoes. Without preface he made them a little speech, in which, pointing at the American flag, he descanted upon the greatness of Lincoln, told how he had liberated millions of slaves in the North and finally announced, in the most matter-of-fact way, that all his slaves were free from that hour.

He proposed to issue at once the necessary official notice of manumission, but it seemed easier for the good man to free thirty or forty slaves than to find pen and ink. Having at last secured these necessary articles and several leaves of paper, torn from a cheap account book, he at once proceeded to set forth his purpose in more or less formal phraseology and very informal spelling. The document was signed and sealed with all due ceremony, and Spalding, Lawford, "Nantucket" and myself were called upon to attach our signatures as witnesses. "Nantucket" regretfully declined for sufficient reasons, a circumstance which our host seemed to feel keenly, as, for some cause, he had fixed upon the quartermaster as being the distinguished head of the whole expedition. The rest of us signed the paper and the old gentleman enclosed it in an envelope addressed to the Minister of Agriculture at Rio de Janeiro. We undertook to mail it, as we afterwards did, in Pará. Shortly after we resumed our way and, with no more than the usual blind luck about the web of channels, reached the *Enterprise* where we encountered the always interesting experience of our own resurrection from the dead. We had been given up as probably lost and our virtues were on everybody's tongue. Within two hours the *Enterprise* had weighed anchor and was standing out between channels in preparation for their escape to sea. So were our virtues.

    I have related this incident at some length as a narrative of the absurdities of two popular superstitions, which have always seemed to me peculiarly objectionable. First, that prior to the Spanish-American War the United States Navy, its officers and men, did little to nothing except to draw their pay. In fact they possessed just the qualities then that afterwards made them so quickly successful at Santiago and Manila and were quite as well prepared then as subsequently to deal with emergencies as they might arise. Second, that prior to our successes against Spain in 1898, the flag of the United States was not particularly respected and her power not sufficiently recognized. In many years of residence in many parts of Spanish America, while in constant contact with the officials and citizens for many nations there and elsewhere, personally, I never saw the time when the flag did not command all the respect that the most thorough going American would desire. I *have* seen the time when our citizens did not receive the consideration they supposed to be their due, but the reasons for this were perfectly obvious and rested chiefly with those citizens themselves.

Poor Spalding, a few years after, was blown up in the course of some torpedo experiments conducted at Newport, Rhode Island. He was a fine type of the genuine sailor man, an officer and a gentleman in the best sense of the words.

With the departure of the *Enterprise* we all relapsed into the hopeless confusion, then attending everything connected with the Madeira and Mamoré railway project. The news from both England and the States was increasingly gloomy. Colonel Church was fighting for the life of his great enterprise in London and all sorts of disappointments and anxieties were being experienced in the United States, by reason of the tying up of the railway company's funds in England. This is a chapter in the history of the undertaking that belongs in other hands. Suffice it to say that each out-going schooner and steamer carried its quota of officials and laborers, who had come down the river and were making their way home. So far as we were concerned, the same process of adjustment and dispersement was carried out. It was a well-nigh heartbreaking experience, but everybody put the best face possible upon it and insisted that there was no intention of abandoning the enterprise, but only a temporary suspension caused by speculative intrigues on the part of powerful and hostile influences in England.

Our own cup of affliction seemed about full, when the *Santa Maria* went to the bottom of the Pará River in forty feet of water. It was a singular performance due to a singular cause. We had taken the large boats *Juno* and *Brazil,* down to the river, towing a couple of schooners out to sea, and had anchored, about seventy miles below the city, to examine a beach nearby, with a view to careening the two large boats and cleaning their bottoms prior to their homeward voyage. The *Santa Maria* had been left at anchor in front of the city, riding to a chain cable which we thought amply long. As we were lunching on the *Brazil,* attention was called to an immense wall of water sweeping in from the sea and extending as far as the eye could reach across the mouth of the Amazon. It was the famous *Pororoca,* or bore, which periodically visits the mouth of that and a few other giant rivers of the world. As it swept by us, it lifted the two boats and sent them forward with such a rush, that the violent strain upon their cables threatened to snap them asunder. In a few minutes the disturbance had passed and all was calm again, as we watched the tossing crest of the water wall disappear in the direction of the city.

When we returned at night, we learned that the *Santa Maria* and the other river craft had been dashed about by the *Pororoca*—intensified near Pará by the narrowness of the channel—and that our launch, having ridden to the limit her cable and her anchor would permit, had simply been engulfed as the great wave rose high above her free board. At that time of the year the current was running at ebb tide with a velocity of over five miles an hour and the river water was little better than solid silt, so heavily was it charged with the enormous scourings of the vast alluvial regions above.

Not a waterman would give us ten *milreis* for the boat, but we decided that we could not afford to lose it. Our little Peruvian deck boy, Pedro, volunteered to dive down, locate the launch and ascertain how she lay. It was no small job, but he was as full of sand as the current, and, after six or eight plunges, was able to give us enough information to confirm our belief that it was worth while attempting to raise her.

To say that we were jeered at by the whole Brazilian population, as our preparations progressed, is to put it mildly, but the fact remains that we did get chains under the boat and did jog her along for one hundred yards or more, until she was in a position where, by the use of lighters, alternately loaded and emptied, we raised her free of the bottom and dragged her into a place of safety. It was the first time anything of the kind had been done in those waters, and her appearance on emerging furnished ample justification concerning the incredulity with which our attempt was regarded. The boat and everything in it was a solid mass, of almost impalpable silt, so fine that the locks of the Winchester rifles, held in the gun racks, were found to be closely packed with mud and even the compass was full of the floury stuff. The work was one that required constant attention day and night, but it was performed with an alacrity and resourcefulness, reflecting high credit upon the men of the expedition force, as well as upon those who were drawn from the crews of the two large tugs. I mention this only as an illustration, possibly unnecessary, of the loyalty which pervaded our entire organization, and is happily felt as the usual in all but the worst managed American enterprises, conducted under novel circumstances and remote corners of the earth. The long existence of Madeira and Mamoré Association itself bears testimony to the prevalence of the same qualities of a larger scale, in the railroad branch of the undertaking.

When the time came for us to disband the expedition and for its members to return to the States, it was with the sincere hope that the difficulties in England and at home might be soon settled, so that our labors could be renewed. This was not to be, and the break-up proved to be final.

Even in those days of deepest disappointment and regret, incidents were occurring which made us forget our troubles. One was the departure of our young interpreter, Ernest Morris, known to newspaper reporters as "The Boy Naturalist of the Amazon."

He had originally gone to South America from Indiana, as an orchid hunter for a wealthy gentleman of Troy, New York, *via* the Wabash, Ohio and Mississippi rivers, over the Gulf of Mexico and the Caribbean Sea, up the Orinoco and down the Rio Negro to the Amazon. He had been quite successful and was employed by us, not only on account of his familiarity with the language of the country, but, also, for his geographical knowledge of a then almost unexplored region. When the end came, he decided to go up the river to collect orchids and trade for rubber on his personal account. To assist him in this, he applied for such portion of our stock of gimcracks and trader's stuff, as we could spare, and was given a free hand in making the selection. This completed, we found he had taken all the playing cards, intended for our own diversion. When asked as to the use he intended to make of this odd addition to his peddler's pack, he replied, "The people on the rivers will think the face cards represent saints and will swap no end of rubber for them." Morris died a short time ago, according to the newspapers, while acting as professor in some Indiana University.

The *Juno* and *Brazil* returned to the States. The three smaller boats were sold to river merchants at Pará. True to her propensities for seeking the lowest level, the *Santa Maria*, while being towed to her destination of one of the upper tributaries, dived to the bottom of the Amazon off Serpa in about the deepest water she could find. Her new owners left her there, and no doubt the fascinating Uyaras turned her into a house boat for their sub-aqueous excursions.

We had about 100 men, all told, engaged in the service and lost none of them, a high tribute, I think, to the skill of Dr. Jack Pennington and the discipline of the respective heads of departments; for the exposure was constant, the vicissitudes many and the stress most wearing. We did not succeed in carrying out our plans, but our failure was not due to any lack of earnest, faithful effort. Every man in a responsible position did his best and a high average of individual accomplishment resulted. Whatever blame attached to our part of the performance fell, where it belonged upon the writer, as head of the expedition.

Subsequent events have abundantly shown that the most serious mistake, committed by Colonel Church and his coadjutors, was the not uncommon one of being about thirty years ahead of the world's progress, a period, according to the late Collis P. Huntingdon, invariably required for the complete development of any great project for the benefit of mankind, attempted under conditions similar to those which confronted us in South America.

XXIX. Results

Not in the clash of steel is found,
For them, the only battle ground.

When P. and T. Collins signed the contract for the construction of the Madeira and Mamoré Railway, they did not consider it necessary to investigate the previous history of the project. They were aware that litigation had followed the abandonment of the enterprise by the Public Works Construction Company but were assured by Franklin B. Gowen that the result had been to make the trust fund in the Bank of England immediately applicable to the purposes for which it had been raised.

Up to May 1879, the expenditures incurred in the execution of their contract amounted to more than five hundred thousand dollars, and they had not received one cent of compensation. They found themselves forced into further unnecessary expenditures in order to protect their private fortunes, involuntarily staked upon the issue of fresh litigation, which had been started in England, though they were neither parties to the suit nor concerned in the disputes that gave rise to it. The sum of eighty thousand pounds was admittedly due them from the railway company. They held certificates for twenty thousand more, which the chief engineer for the company refused to sign, and they had done a large amount of extra work, on grading and laying track over temporary line, for which, under their contract, they could claim no compensation.

Resident Engineer Nichols said under oath afterward that the work on the first six miles south of San Antonio was fully six times as great as the average work per mile, estimated for the entire railway. The surveys and estimates confirmed this statement, and the location of the first five miles out of San Antonio, under conditions imposed by contract as to grade and curvature, had proved to be a more difficult problem than the location of the remaining sixty-two miles to the Caldeirão do Inferno.

"The beginning is half the whole work" is an aphorism preeminently applicable to projects executed by civil engineers under novel conditions and in places remote from civilization. The preliminary experience, though invaluable later, is very costly and discouraging at the time it is acquired. The transportation of men and materials to the field of operations is generally in itself sufficiently expensive to render the introductory labors unprofitable. The Public Works Construction Company had received fifty thousand pounds in advance and had forced Colonel Church into promising forty-five thousand more for work and material that were practically worthless. With impaired credit and unpaid labor, P. and T. Collins had successfully inaugurated a most difficult undertaking, achieved results, which under the circumstances, were surprising, and yet were unable to obtain amounts confessedly due them by the terms of a contract. They had shipped to San Antonio tremendous quantities of provisions, tools, mechanical appliances, rolling stock, and railroad iron and had given out numerous contracts that could not be repudiated.

The actual work performed may be briefly summarized as follows:

- Aggregate length of all lines cut and surveyed through the forest: 320 miles;
- A projected location made, extending from San Antonio to the Caldeirão do Inferno for a distance of sixty-seven miles;
- Location of railway finally established on the ground and accepted by the railway company: thirty-six miles;
- Additional information marked on the ground but not finally accepted by the railway company: four miles;
- Right-of-way, one hundred feet wide, cleared of timber and brush for twenty-five miles;
- Grading completed or in progress for seven miles;
- Construction train running out of San Antonio for 3.9 miles;
- Permanent main track laid for 2.2 miles;
- Temporary track laid for 1.7 miles;
- Telegraph poles erected for 2.8 miles;
- Telegraph line completed for 1.6 miles;
- Thirty thousand cross-ties made;
- Seven buildings erected at San Antonio;
- Nine trestles constructed (aggregate length of trestles was 1,453 feet).

Besides making free use of the facilities for river transportation offered by the Amazon Navigation Company, the contractors had chartered three ocean steamships and seven sailing vessels in the United States. They also had, for some time, the exclusive use of the largest passenger steamer on the Amazon and kept the two steam tugs of Mackie, Scott, and Co., as well as their own little steamer, the *Carlo de Lima*, constantly occupied.

The total number of persons sent from the United States to San Antonio is shown in the following table:

Ship	Number of Persons
Mercedita	220
Metropolis	222
City of Richmond	467
Juno and *Brazil*	30
Carlo de Lima	1
Enlisted at Santarem	1
Total	941

As none of the passengers on the *Metropolis* ever reached their destination by that vessel, only 719 of the entire number of persons dispatched from the United States, arrived at San Antonio, and of these, about half a dozen were women.

About two hundred Bolivian Indians and four hundred men from the Province of Ceara were employed on the work, but the total working force never exceeded one thousand men at any one time and only for a brief period approached the maximum.

No record of mortality among the natives was ever kept,

though many are known to have perished. The death rate among the men from Ceara, owing to the filthy habits, extreme poverty, and ignorance regarding the country and climate, was exceptionally high. Almost nothing is known about the fate of many laborers from the United States who attempted to reach Pará in canoes and on rafts. A number of them are believed to have died either on their way to or after their arrival in that city. Not a few men died after returning to their homes in consequence of disease acquired in Brazil, but there is no means of ascertaining the number of such deaths. The mortality among employees from the United States, so far as positively known, was as follows:

Perished in the wreck of the *Metropolis*	80
Died on the *City of Richmond*	1
Died at San Antonio	56
Perished in an attempt to reach Bolivia overland	75
Killed by savages above San Antonio	1
Died on the *Brazil*	1
Died on the *Theotonio*	5
Died at Manáos	1
Died at Pará (so far as positively known)	1
Total	221

Killed or Wounded by Savages

Americans killed	1
Bolivian Indians killed	2
Americans wounded	2
Bolivian Indians wounded	3
Total	8

Of the 941 persons who left this country for San Antonio in 1878, it is absolutely certain that at least between 22 and 23 percent lost their lives. The significance of this statement will be more apparent when we reflect that the mortality resulting from accident, wounds, and disease among the Union soldiers during our four years of bloody civil war was only 10.5 percent. This comparison carries with it a lesson for those whom music, gilt braid, and brass buttons have hypnotized into believing that only in military life is there to be found any adequate field for the display of physical courage and endurance.

In an oration delivered at Harvard University a few years ago, William Everett said, "There are more heroic and sacrificial acts going on in the works of peace every day than the brazen throat of war could proclaim in a twelvemonth." Let the reader who doubts this assertion compare the statistics just given with those of our late war with Spain. During that conflict, the death rate in the American army, from all causes, was *only 1 percent*, or about half that frequently observed in a single year among the peaceful residents of our large cities. Only three employees of P. and T. Collins were killed by savages, but even this small number gives a death rate more than double that due to injuries inflicted by Spaniards upon the American troops in 1898. At the end of a seven-week campaign in Cuba, General Shafter's army, notorious in war, was so completely demoralized and conquered by malarial fever that 90 percent of the soldiers were unfit for duty and many of the officers signed a round-robin urging the government to bring them home. In the face of much greater proportionate mortality, it required a year and a half to starve the unpaid railroad men of 1878 into a complete abandonment of the enterprise in which they were engaged. Even then, the rear guard of determined men only left San Antonio when compelled to do so on positive orders from their superiors at home.

An infatuated populace never tires of heaping political and military honors, as well as pecuniary rewards, upon the gilded idols who have forced us to dissipate our national wealth and energies among Malays of the East and to alienate the sympathies of a race that still controls half the Western Hemisphere. For years to come, army promotion and retirement from active service upon liberal salaries and pensions will continue to follow those who were engaged in the opera bouffe warfare of 1898. Time has mitigated the humiliating spectacle presented by groups of gushing girls waiting at railway stations to perform previously unheard-of feats of osculation upon a young naval hero who made an abortive attempt to bottle the Spanish fleet in the harbor of Santiago, but we may even yet see a general at the head of our armies who obtained his military training in a medical school, fought in Cuba the only trouser-wearing foes he ever encountered, and reached the climax of a brief but brilliant military career at the indiscriminate massacre of men, women, and children in the crater of Mount Dajo.

The railroad men, who, in 1878, risked life and health in a vain endeavor to develop, by peaceful methods, vastly greater latent wealth than any which exists in the Philippines, have long since been forgotten. There was nothing of a spectacular character to attract public attention in their prosaic attempt to establish direct commercial intercourse between this country and the interior of South America. A minute fraction of the money wasted in and in consequence of the war with Spain would have changed their failure into a brilliant success, would have established a permanent community of interest between ourselves and the nations south of us, would have given us a commanding influence in the commerce of the two Americas, and would have caused the consignment of the Monroe Doctrine to the rubbish heap, as a useless expression of the proposed attitude of the United States under certain conditions, which could not by any possibility occur.

The actors in the drama of 1878 had none of the requisites essential to enduring fame. They wore no gaudy uniforms, carried no flaunting battle flags, required no strains of martial music to make them forget danger in the performance of duty, and had no retinue of newspaper correspondents to chronicle the death of every mule that fell victim to their accurate gunfire.

Today, many of them are still to be found employed in positions of great responsibility, on railways and public works in the United States, South America, and Mexico, but, as one by one, they pass to their eternal campground, no American flag drapes their coffins and no farewell volley is fired over their graves. Such honors are the exclusive prerogative of those who consecrate themselves to the destruction of human life and the dissipation of human energy.

XXX. The Litigation in England

*Where suits are traversed, and so little won
That he who conquers is at last undone.*
—Dryden

When, on December 11, 1874, Sir George Jessell, Master of Rolls, handed down a decree liberating the fund on deposit in the Bank of England and removing all doubt regarding its applicability to the construction of the Madeira and Mamoré Railway, superabundant caution and consideration for possible but improbable claims, which might be urged in behalf of a party outside the court's jurisdiction, led him to insert a qualifying clause, stating that his decision was made "without prejudice to any claim of the Republic of Bolivia." In explanation, he said, "I do not know what the claim is, and I do not know that there is a claim. In fact, I may say that at present I am satisfied on the present papers there is none, and there must be further papers and some further evidence to convince me of it."

This decree followed closely the termination of the suit instituted by the Bolivian bondholders, when the plaintiffs were curtly informed that "there is no equity in this bill, and therefore, I am bound to dismiss it."

Early in 1877, the Bondholders' Committee in England sent a Mr. Richard Reader Harris to La Paz with authority to expend twenty thousand pounds in an effort to secure the annulment of Colonel Church's Bolivian concession, an assignment to the bondholders of all claims Bolivia might have on the trust fund, and the right in her name to prosecute such claims in the English courts.

The negotiations continued until December 1, 1877, when Mr. Harris finally accomplished, by methods which, to say the least, were open to suspicion, what Colonel Church justly characterized as "his unsavory mission."

The Bolivian Congress then ratified an agreement by which all the rights, privileges, and obligations in Bolivia, connected with the Bolivian loan, were transferred to the bondholders, with power, in the name of Bolivia, to revoke Colonel Church's concession, abandon his projects, and, so far as the courts would permit, appropriate the trust fund.

In a circular issued by Mr. A. W. Ray, the chairman of the Bondholders' Committee, under date of March 1, 1878, it was stated that:

> The fund in the Bank of England, which we seek to have distributed among the bondholders, now amounts to £743,000, is equal to about £45 per £100 bond. ... It is impossible to construct the railway and legal proceedings are now in progress with the object of effectually preventing the waste of the fund in any attempts to do so. ... The new contractors have made

a formal application to the court for a payment from the fund which has been refused.

It is, therefore, apparent that, at a time when the American engineers had only been ten days at San Antonio, when the wreck of the *Metropolis* had brought them in sight of starvation, and when the *City of Richmond* was still at sea hastening to their relief, the British bondholders had already tied up the fund upon which P. and T. Collins depended for the payment of salaries and expenses and were declaring the construction of the railway impossible.

On March 2, 1878, a representative of the bondholders began a suit in the Chancery Division of the High Court of Justice, praying that the trust fund might be divided *pro rata*, according to holdings, among them and that the trustees might be restrained from making further payments for work on the railway. The defendants to this suit were the National Bolivian Navigation Company; the Madeira and Mamoré Railway Company, Limited; the Public Works Construction Company; the trustees of the trust fund; and a single bondholder, who objected to the course of the majority. The case did not come to trial until April 23, 1879.

Divested of legal verbiage, repetition, and amplification, the claim of the plaintiff was substantially as follows:

(1) That the Bolivian concession had been obtained by fraudulent and untruthful representations made by Colonel Church;

(2) That the Brazilian concession was made to Colonel Church personally, subject to the law of the empire, and had expired by limitation;

(3) That the Public Works Construction Company had not earned the ninety-five thousand pounds paid or promised to it and that such actual or proposed payments could only be made by a misappropriation of the trust fund;

(4) That all subsequent contracts for construction of the railway had been executed without the consent or concurrence of the bondholders;

(5) That the length of the proposed railway was unknown; the statement of claim contained further allegations in these words:

> The country through which the railway must necessarily pass is a dense and almost impenetrable jungle, full of swamps and lagoons and intersected by deep ravines and in many parts flooded by inundations to a great depth during a great portion of the year. The climate of the country is also most pestiferous and deadly. It is impossible for white laborers to endure the climate for more than a few weeks. There are, moreover, savage tribes, whose attacks prevent any continuous progress with the works. The difficulties of constructing the railway, owing to the nature of the country through which it is to pass and other circumstances, are practically insuperable … and the railway cannot be made without great peril to and sacrifice of human life.

(6) That the cost of the railway, under the contract with P. and T. Collins, would far exceed the amount of the trust fund;

(7) That Colonel Church had represented in the prospectus issued to investors, as well as to the Bolivian government, that the railway could be constructed in two years, that this time had already expired, and that no one could state definitely when the road could be completed;

(8) That interest on the bonds had not been paid since June 1875, and that interest previously paid came out of money subscribed by the bondholders;

(9) That no bonds had been redeemed, and that the sinking fund requirements, stated in the prospectus, had been disregarded;

(10) That neither the navigation nor the railway company had any money invested in Colonel Church's projects, and that neither had any capital except that derived from the Bolivian loan;

(11) That the prospective earnings of the two companies would never pay operating expenses;

(12) That the security offered in the prospectus to bondholders was "illusory and valueless";

(13) That lack of funds made it impossible for Colonel Church to comply with the terms of his Bolivian concession;

(14) That fraud made the concession voidable by Bolivia, and that Bolivia had completely revoked and annulled it;

(15) That P. and T. Collins had undertaken to build the railway with a full knowledge that neither Bolivia nor the bondholders desired it, and that both objected to the use of the trust fund for that purpose;

(16) That P. and T. Collins, under their contract, could abandon the work, if the prescribed payments were not made, and that, as the trust fund was insufficient, the only possible result would be an abortive attempt to construct a railway, through a country without population or resources, which would be useless for practical purposes.

In a counterstatement, the defendants gave the entire history of Colonel Church's projects and denied most of the plaintiff's allegations. The case was heard by Mr. Justice Fry, and Judah P. Benjamin, formerly attorney-general for the Southern Confederacy, was the leading counsel for the defense. At the very outset, Mr. Benjamin informed his opponents that he wished to be perfectly fair and, for that reason, gave them notice that it would be well to introduce *all* their evidence in opening the case, as there would be no rebutting testimony, and he proposed to ask for a nonsuit on evidence presented by the plaintiff.

Some months previous to the trial, the bondholders had sent Mr. Harris and William Cunningham, a mechanical engineer, to San Antonio, in order that they might be able to testify regarding the progress made in building the railway. Harris was armed with an order from the master of rolls authorizing him to make an examination of the work, and so much importance was attached to the evidence these two men were expected to give that the bondholders had secured a postponement of the trial until their return. They had gone together from England to Pernambuco and thence to Pará, where it seems, for some unknown reason, over the railroad; *tomorrow*, they would take great pleasure in getting out all maps, plans and estimates to explain everything he cared to know, but until *tomorrow*, they were really too busy to talk much. Finally, an hour and a half before sunset on Sunday, they consented to gratify his curiosity. Steam was raised on a locomotive, and he was taken about three miles outside of San Antonio. On the way, much valuable information was given him about the savages who inhabited the country, and many were the bloodcurdling tales of their atrocities related to satisfy his inquisitive mind. At the end of a high trestle, the locomotive came to a stop and, with only an hour of daylight ahead of him, Cunningham was told that, owing to repairs in progress on the trestle, it was unsafe for the locomotive to cross but that they would be glad to wait for him while he walked to the end of the track. He did walk a short distance but never let the locomotive get out of his sight. The forest had already become dark and gloomy, and his imagination had been rendered abnormally active by the terrible tales of cannibal orgies with which his head had been filled. He hesitated, and at last, his personal courage yielded to a practical consideration of the consternation and legal complications that would inevitably follow should the English bondholders learn that their most important witness had gone where nothing short of a writ *de ventre inspiciendo*, served on South American savages, could discover him.

He returned to San Antonio, sailed for Pará the next day, and in due time, after traveling about twelve thousand miles, presented himself as a witness before the master of rolls.

As may be inferred from the foregoing explanation, his evidence proved disappointing. He did not see anyone working on the railway—a fact not surprising since the day was Sunday. He did not know how far the line was cleared of timber, because he was not within twenty miles of the end of the clearing. He did not see the end of the track and could not state from personal knowledge how far it extended. Asked why he had not walked to the end, he replied, "I was not going to walk up there by myself."

At this point, Justice Fry could no longer repress his feeling of astonishment at such testimony from a witness, whom he had postponed the case to hear, and he exclaimed, "I confess this is the most extraordinary evidence I ever heard."

Mr. Day, attorney for the plaintiffs, added, "It is an extraordinary place to go to. I understand from Colonel Church's book that it was not safe to venture on account of the natives."

The bondholders were little more successful with their other witnesses and utterly failed to substantiate their claims. No testimony was offered by the defendants. Mr. Benjamin claimed a nonsuit, and Justice Fry, after a review of the argument made on behalf of the bondholders, closed with these words: "That argument, in my judgment, fails, and therefore, I accede to the suggestion of the defendant that the plaintiffs have shown no case in their opening and the result is that I dismiss the action with costs."

The case was then carried to the Court of Appeals, where it was practically retried by the Lords Justices James, Brett, and Cotton. Voluminous testimony was offered by the defense. Among the witnesses were Colonel Church, Thomas Collins, O. F. Nichols, C. S. d'Invilliers, and a number of others. Unfortunately, the evidence so painfully lacking at the first trial was abundantly supplied at the second, and no witness for the bondholders could have been more useful to them than Mr. Thomas Collins, who, in his bluff, hearty way, told what he believed to be the truth without, apparently, any regard to its effect on his own interests. He made the fatal admission that it would cost more than one million pounds to build the entire railway, an amount that greatly exceeded the trust fund, and stated that he relied upon the Brazilian guarantee to supply the deficiency. The result of the trial was a reversal of the judgment given by Justice Fry and an order directing the distribution of the trust fund among the bondholders ratably according to holdings.

The litigation was accompanied and followed by many venomous newspaper attacks, which appeared in the principal cities of England, the United States, and South America. Sometimes, Colonel Church was the victim selected, but in the United States, the English courts and English judges came in for their full share of abuse. A most careful study of the evidence given and the arguments made in the Court of Appeals, supplemented by the reading of many newspaper articles published at the time, convinces the writer that there never was any sufficient ground for imputing improper motives, intentional injustice, or dishonorable conduct either to Colonel Church or to the lords justices of appeal. The justices did not even discuss or consider the possibility of constructing the Madeira and Mamoré Railway as an engineering problem. Neither did they attempt to decide whether the railroad, if constructed, would be a financial success and accomplish all the beneficent results claimed by Colonel Church.

The decision was substantially this, that *as a legal and financial proposition, it was impracticable to attempt building the entire railway with any funds then available for the purpose.*

An analysis of the three masterly opinions handed down shows that the reasoning of the judges rested on two premises. If these premises were false, the opinions are worthless and fall to pieces of their own weight, but if true, we are forced by inexorable logic to the same conclusion reached by the lords justices of appeal. These fundamental facts or alleged facts were:

(1) That the Bolivian concession had been revoked by competent authority;

(2) That it was impossible, with the trust fund alone, to build the entire railway.

In the trial, the defendants made no attempt to deny or dispute the first statement.

Whether correctly or not, on the witness stand, Mr. Thomas Collins had distinctly admitted the second statement to be true. He may or may not have been wrong, but the defendants could not dispute the evidence of their own witnesses.

After some trouble and further heavy expense, permission

was obtained to bring the case before the highest judicial tribunal in Great Britain and, a little later, the House of Lords, acting as a court of appeal, confirmed the decision of the court below.

While earnestly maintaining the impartiality and fairness of the English courts, it would be very far from the truth for the writer to assert that the result of the litigation did not work great hardship to many persons, who, through ignorance or bad legal advice, had identified themselves with an enterprise of far-reaching importance. Neither can the fact be concealed that the principal beneficiaries of the litigation were an unscrupulous body of commercial pirates, who had purchased Bolivian bonds at a mere nominal price and with the deliberate purpose of wrecking the whole scheme, which Colonel Church had matured after many years of laborious effort.

The law student who defined a court as "a place where injustice is judiciously administered" was not altogether wrong. In law, as in medicine, the most approved remedies often have deplorable aftereffects and no human tribunal ever has been or ever will be capable of solving all the complex problems that come before it without damage to the interests of persons innocent of any intentional wrongdoing.

The direct consequences of the legal decisions noted above may be briefly stated as follows:

Although the legal contest had been entirely over the railway trust fund, the final decision applied with equal force to all the then existing proceeds of the Bolivian loan, including the amount set aside for use of the National Bolivian Navigation Company. These two funds originally amounted to £708,228, but they had been so judiciously invested by Colonel Church that, aside from the accumulative interest, a profit of more than £150,000 resulted; so that the net proceeds of the Bolivian loan, notwithstanding expenditures, ultimately amounted to more than £850,000.

The entire cost of wrecking the enterprise, including £27,000 paid to Richard Reader Harris for his services in securing a cancellation of Colonel Church's Bolivian concession, was £59,632. The balance of the fund was distributed among the bondholders.

The motive, which led London speculators to purchase Bolivian bonds in the open market at 16 percent of their face value, becomes clearly apparent when we see that funds of the railway and navigation company were sufficient to pay 52 percent of the face value of every bond issued.

P. and T. Collins never received any compensation for their expenditures and were driven into bankruptcy. Their losses are supposed to have amounted to $800,000. The devoted wife of Mr. Thomas Collins, who had stood loyally at his side throughout the whole trouble, when the litigation in London terminated, had to be confined in an asylum for the insane, where she died not long after.

For a number of years, P. and T. Collins were unable to pay large sums due their employees and others, but they never regarded their own misfortunes as a sufficient excuse for involving their subordinates in the calamity that had overtaken them, and, in later years, when fortune had smiled upon their labors in more productive fields, they divided their surplus earnings among those who had legitimate claims against them. They were never able to fully discharge their financial obligations, but they made an earnest and honest effort to do so. When, a few years ago, death claimed both members of the firm, no one more sincerely mourned their loss than those who participated with them in the disastrous expedition of 1878.

Colonel Church, finding the fruits of his unremitting labors for eleven years hopelessly dissipated, went to Rio de Janeiro and had the Brazilian concession revoked, in order that enemies of the enterprise might in no way benefit by what he had accomplished in negotiations with the emperor.

No party to the litigation had so much at stake upon the successful execution of Colonel Church's plans as the Republic of Bolivia. It is true that circumstances beyond his control, which he could by no possibility have foreseen, had caused delay and expense never anticipated, but had the Bolivian government stood loyally by him and by their own interests in the hour of adversity, maintained Bolivian credit, and paid the interest due on Bolivian bonds, there would have been no possible ground for litigation. But the same policy of secret intrigue, the same disregard of national honor, and the same shortsighted statesmanship, which during the administration of President Belzu had caused Bolivia to break treaties and answer the protest of foreign ministers by expelling them from the country and brought the British Foreign Office to declare that she could no longer be regarded as a civilized nation, were once more in supreme control of her destinies.

Bolivia, "the altar upon which the first blood was shed for liberty and the land where the last tyrant perished," in the long struggle for South American independence; Bolivia, with all her glorious past and her grand possibilities for the future;

Bolivia, with undeveloped resources, which should have made her the dominant force in the trade and commerce of South America, was once more the victim of a conscienceless usurper, who was faithless to her highest interests.

Colonel Church gained every suit in the English courts, which he defended, prior to the revocation of his Bolivian concession and, even after that revocation, obtained a favorable decision from the master of rolls on *ex parte* evidence presented by his opponents, but Bolivia, by her own action, had discredited her own financial agent, destroyed the only existing security for repayment of the Bolivian loan, made it impossible for Colonel Church to defend his projects against the avarice of London speculators, and postponed for the far-distant future the utilization of the Amazon and Madeira Rivers, as a great natural highway for Bolivian commerce.

To complete the wreck of hopes long and fondly cherished by almost every distinguished statesman she had produced up to that time, Bolivia became entangled in an intrigue and secret alliance with Peru, violated her treaty obligations to Chile, and, at the very time the bondholders' suit was still pending in the Court of Appeals, became involved in war with the last mentioned power, which finally resulted in the loss of all her territory on the Pacific coast.

Isolated by her own perfidy, she had since been compelled to pay tribute to surrounding nations for the privilege of carrying on foreign commerce over their territory and may, at any whim of theirs, lose even the small consolation their charity or selfish interest now leads them to offer.

XXXI. The Recent Revival of Colonel Church's Project

In the field of thought,
nothing save the chaff perishes.
—W. Fraser

Such a record of disaster as the preceding pages contain would naturally lead one to expect the abandonment for all time of Colonel Church's plans and projects, but there was then at the head of the Brazilian government one of the wisest, most enlightened, and most far-seeing statesmen who ever guided the destinies of the empire, one whose faith in the Madeira and Mamoré enterprise could never be shaken and who never ceased to express admiration for and confidence in the indomitable will and unceasing energy of its projector. That man was Dom Pedro II, Emperor of Brazil. He looked with disfavor upon the attitude assumed by Bolivia and deplored the resulting damage to commercial progress throughout the territory drained by the Amazon.

The Chilean war had taught Bolivia some valuable lessons and had produced within her borders an era of reform and wise government. Deprived of her seacoast, she began to cultivate the friendship of Brazil and, early in 1882, on her own initiative, entered into an agreement with that empire that promised to yield her certain trade advantages based upon the conceded importance of constructing the Madeira and Mamoré Railway. Mindful of Bolivia's faithlessness to that enterprise in the past, Brazil insisted upon inserting in the agreement a clause which reads, *"the present treaty will come into force when the railway may be opened to traffic and for fifty years thereafter."*

The frequently repeated allegations made in England, to the effect that the construction of the proposed railway was impossible or impracticable, had so disturbed public confidence that the emperor decided to establish, beyond question, the truth or falsity of all such statements, and it was due to his insistence that in 1882, a large corps of engineers was sent to verify all previous surveys and make an independent estimate of the cost of building the railway.

Dr. C. A. Morsing, an engineer of much ability and force, was in charge of the expedition. Dr. Morsing was born in Sweden, educated in the United States, and naturalized in Brazil. Dr. Morsing's plans were very satisfactory, but he seems to have proved unequal to the emergencies he had to encounter in the field and, so far as the writer can learn, accomplished absolutely nothing beyond taking some photographic views in order to show the actual condition of affairs at and above San Antonio.

In the following year, Senhor Julio Pinkas, an Austrian engineer officer, who had been a subordinate of Dr. Morsing and enjoyed much local prestige, was placed at the head of a commission of fifteen members with authority to employ all assistance necessary to complete the railway examination.

The commission sailed from Rio de Janeiro on January 10, 1883, and from notes made by two eyewitnesses of the embarkation, we present the following quotations. One of these observers writes:

> I can yet see the swarm of dramatically dressed young engineers, with patent leather boots, spurs, revolvers, field glasses, and fetching pith helmets distinctly in evidence, while parading in and out of the fashionable cafés of Rio, prior to embarking on their patriotic mission of teaching *Os Yankis* a lesson in railroad building.

The other says:

> We went on board the steam packet *Espirito Santo* on which they were to sail, with considerable curiosity and the two of us, who were relicts of the American expedition, stood aloof and watched the proceedings with great and not unnatural interest. Both of us well—too well—remembered the departure of the various sections of our expedition from the United States and, when we saw these carefully selected scions of the nobility and aristocracy of Brazil falling upon each other's necks and any—or every—neck within reach, while weeping and wailing out their adieus with floods of tears, we made up our minds that the kind of material required to resurvey our old tramping ground was not on hand. The scene on board beggared description and the howling dervishes,

silenced by envy, would have given up their profession on the spot.

Senhor Pinkas and his men arrived at San Antonio on March 19, 1883, and he himself says in his official report that the consternation that prevailed among his young men on the following day was singularly contrasted with the contentment they had exhibited during the voyage.[28]

On April 13, a field corps of twenty-nine men began surveying, at the rate of a quarter of a mile per day, the railroad track laid by P. and T. Collins, but apparently, they never went as far as the locomotive, *Colonel Church,* had frequently gone less than five years before. Cases of sickness became frequent; one man died, and three graveyards in sight were unpleasant reminders of the fate that had previously overtaken many laborers in the same field. The sick were sent as fast as possible to Manáos, and this process so crippled the party that the commission, which disembarked festive and happy on March 19, two months later was in a completely disorganized condition.[29]

It soon became apparent that it would require at least two years for the Brazilian survey party to retrace and verify the railway location made by the American contractors in nine months, and Senhor Pinkas, with discernment which does

---

[28] "No dia seguinte partiu o vapor que nos tinha levado aquellas plagas e singularmente contrastu a consternação, alias natural, que então se pintou no semblante dos nossos jovens com o contentamento que manifestaram durante a viagem," Pinkas.

[29] "A commissão, que 19 de Março tinha desembarcado festive e alegre, achava-se, pois a 19 de Maio, dous mezes depois, em plena debandada," Pinkas.

him credit, began to see previously unsuspected merits in the work of his predecessors. He became convinced that the process of verification could be carried on more expeditiously and more satisfactorily in the salubrious climate of Manáos than in the forests above San Antonio. He even suspected that with such verification, it might be possible to accept, unmodified, the entire American location.

Accordingly, work at San Antonio was abandoned, and the field party dispatched to Las Pedras on the Jaci Paraná, where Senhor Pinkas arrived in person on July 11. His report mentions as something extraordinary that, nine days later, twenty of his forty-five men were invalids. Commenting upon this terrible state of affairs, he exclaims, "How much longer can we stand it?" A few minutes later, he discovers a remarkable coincidence and adds, "Fortunately this concluded the verification of this part of the work."

He plainly says that troubles incurable by medicine increased the demoralization of the party and either brought on or aggravated the attack of fever.[30] At this time, fond parents of the young engineers, who had learned of the one death that occurred on the corps, thoroughly alarmed for the safety of their precious offspring, were writing and encouraging the young men to resign and come home, but to their credit, it can be said they did not at once yield to such seductive influences.

---

[30] "Dos 45 individuos de que se compõe o pessoal hoje 20 com parte de doente. Entram nessas molestias, para muitos, causas moraes, que, incuravels pela medicina, aggravam—se e as mais das vezes determinam accessos de febre. Quanto tempo resistiremos ainda? Ponderemos permanecer acqui até que venha reforço on auxilio da côrte? *Felizmente está concluida a verificaço desta parte*," Pinkas.

August 23 found our hero at the Caldeirão do Inferno, where he mentions having seen the ruins of a house formerly occupied by the *deceased* Ignacio Arauz. On the thirty-first, he reached the Falls of Girão and on the following day was himself stricken with fever. Under the influence of morphine and quinine, he discovered himself and his party four days later at San Antonio, hardly knowing how he got there. Compelled to wait for a steamer, Senhor Pinkas employed his lucid moments in an examination and *verification* of the Collins plans, in telegraphing to the government his approval of the entire Collins location and in noting the condition of his men.

Under the date of August 15, he writes, *"All are sick. Not a single person in good health."*

On August 19, exactly five months after their arrival, the entire party sailed for Manáos, taking with them all the plans and notes left by P. and T. Collins in the office at San Antonio. One engineer and three men died during the process of verification. The net results were a few useless astronomical observations and the acceptance by the Brazilian government of the entire projected railway location made by the American contractors, extending from San Antonio to the Caldeirão do Inferno, a distance of 106 kilometers. Of the entire route, thus *approved* and *verified*, neither Pinkas nor any of his men ever saw a hundred kilometers. Some of it never existed except on paper and much of it, actually traced on the ground, had never been finally accepted and approved either by the engineers of P. and T. Collins or by the resident engineer for the Madeira and Mamoré Railway Company. The illustrious Pinkas had, therefore, overdone the verification business and had given a more unqualified endorsement to the work of his predecessors than anyone had ever claimed for it.

In the following year, Senhor Pinkas was more successful. With the unlimited resources of the Brazilian government to draw upon, he returned to San Antonio March 17, 1884, and with five large field parties continued the Collins survey from the Caldeirão do Inferno to Guajará-Merim, where the last stake was driven on September 5, 1884. The work was far from satisfactory, the best part of it being a line surveyed across the great bend in the Madeira River above Tres Irmãos. No attempt was made to test the possibility of shortening the line in many other places. Of the entire distance, 329 kilometers, 106 were located by P. and T. Collins, 123 were in shortcuts across bends in the river, and for 100 kilometers more, the line simply followed the riverbank without regard to cost, distance, or topography. Upon such data, Senhor Pinkas claimed the construction of the railway to be perfectly feasible and estimated the cost at £873,671, including equipment and an allowance of 10 percent for contingencies.

Later, Colonel Church is said to have made a new proposal to the emperor of Brazil for the construction of the projected railway. Mr. Mackie, who then represented Colonel Church at Rio, says that Dom Pedro "supported it earnestly and often talked to me, as Colonel Church's attorney, of his great admiration for and confidence in that indomitable man, but political sentiment was then averse to letting any foreigners build what Brazilians regarded as the key to the heart of South America. The shadow of the Republic was already spreading over Brazil and nothing definite resulted."

Thus it happened that for many years, the Madeira and

Mamoré project has been held in abeyance. The quinine in the forests of Bolivia has been exhausted, but the rubber trade on the tributaries of the Amazon became so valuable as to cause a Spanish adventurer to proclaim the existence of the "Independent State of Acre," embracing disputed territory bordering on Brazil, Bolivia, and Peru, of much greater extent than the whole of New England.

The obliteration of this ephemeral creation almost precipitated a bloody conflict between the three contiguous nationalities over their respective claims to the rubber forests, which constituted all there was of the evanescent republic. Happily, these claims were all successfully adjusted by the Treaty of Petrópolis, signed in 1903 and ratified in the following year.

Under this treaty, in exchange for territorial concessions, Brazil agreed to build the Madeira and Mamoré Railway within four years and to pay Bolivia two million pounds, which the latter accepted *"with the intention of applying it principally to the construction of railways and other works tending to better communications and to develop commerce between the two countries."*

Recently, New York bankers with European connections are reported to have undertaken to provide Bolivia with a network of railroads costing twenty million dollars, the key to which is found in our almost-forgotten scheme of river transportation, and as we go to press, the newspapers announce that the government of Brazil has entered into a contract that assures the ultimate execution of the wide plans made by Colonel Church forty years ago and the immediate construction of the Madeira and Mamoré Railway.

Dawson, writing in 1904, says of Bolivia:

> Her great resources can never be profitably utilized until a practical outlet to the sea has been found. ... The explorations of Heath on the upper tributaries of the Madeira resulted in discoveries which may ultimately enable Bolivia to utilize the magnificent fertile plain just at the foot of the table land, but so far well-nigh as inaccessible as the South Pole. *Broad and navigable rivers meander through this vast region needing only the construction of a railway around the Madeira rapids to communicate with the Amazon and the Atlantic.*

Since 1878, notwithstanding the obstructions above San Antonio, many small steamers have been placed on the Bolivian tributaries of the Madeira and today are navigating for many hundreds of miles the waters of the Beni, Madre de Dios, and Mamoré, but the Indian canoe is still used to transport freight and passengers between Guajará-Merim and San Antonio.

Probably no one now living can more than faintly realize the wonderful results sure to follow the establishment of any adequate system of transportation between these two points. It is a well-authenticated fact that some years ago, a rubber merchant of Trinidad, rather than endure the fatigue and hardships of an overland journey to La Paz, distant about 285 miles in an air line, descended the Mamoré, Madeira, and Amazon Rivers to Pará, thence went by sea to Colon, by rail over the Isthmus of Panama, by sea to Mollendo in Peru, by rail to a steamer on Lake Titicaca and finally by a few miles of travel over a good road reached his destination.

This illustrates the importance of Bolivian rivers merely as a means of internal communication under conditions now existing. No railroad from the Pacific coast is ever likely to be extended over the Andes into the lowlands of northern and eastern Bolivia until the wealth and population of the country have increased prodigiously. What the effect of making this vast and fertile region easily accessible from the Atlantic coast will be, no man can foretell, but two things are certain. A greater commercial development will follow than any ever predicted by Colonel Church, and the nation which controls the Madeira and Mamoré Railway is sure to exert a far-reaching influence over the trade and commerce of the two Americas.

Appendix

Observed Elevation of Water Surface in the Madeira River above the Low Water of 1878–1879

March 1, 1878 ... 43.0 feet
September 6, 1878 ... 3.9 feet
September 14, 1878 ... 1.7 feet
September 25, 1878 ... 0.0 feet
October 1, 1878 ... 1.7 feet
October 15, 1878 ... 7.3 feet
November 1, 1878 ... 10.6 feet
November 15, 1878 ... 18.3 feet
December 2, 1878 ... 18.8 feet
December 12, 1878 ... 23.2 feet
January 2, 1879 ... 22.5 feet
January 13, 1879 ... 21.8 feet

January 30, 1879 ... 27.2 feet
February 13, 1879 ... 29.0 feet
March 3, 1879 ... 37.0 feet
March 13, 1879 ... 40.5 feet
April 3, 1879 ... 37.2 feet
April 14, 1879 ... 38.6 feet
May 1, 1879 ... 31.7 feet
May 15, 1879 ... 26.4 feet
May 29, 1879 ... 15.9 feet
June 16, 1879 ... 10.7 feet
June 30, 1879 ... 11.6 feet
July 7, 1879 ... 9.9 feet

Mean rise and fall of river: 41.75 feet

Temperature and Rainfall at San Antonio

Note—The temperature is given in degrees of Fahrenheit thermometer, rainfall in inches	Average daily maximum temperature	Average daily minimum temperature	Average daily variation of temperature	Total rainfall
June 1878	90.4	70.1	20.3	0.436
July 1878	91.2	68.4	22.8	0.010
August 1878	92.2	67.5	24.7	0.535
September 1878	90.7	71.4	19.3	1.395
October 1878	90.3	72.2	18.1	8.265
November 1878	89.0	72.7	16.3	7.720
December 1878	88.8	74.6	14.2	11.311
January 1879	88.0	72.4	15.6	15.200
February 1879	88.2	72.6	15.6	10.245
March 1879	87.0	71.9	15.1	18.412
April 1879	88.3	72.1	16.2	11.399
May 1879	90.6	70.9	19.7	0.853

Total annual rainfall: 83.781

Note—The annual rainfall, as determined by the English in 1873, was 81.52 inches.

The Navigation of the Madeira River

Without doubt, much of the trouble experienced in bringing the *Mercedita* up the Madeira to San Antonio was due, not so much to insufficient water in the channel, as to the nonexistence of charts showing where that channel was situated. In February and March 1878, the water in the Madeira must have been about thirty feet above its lowest stage, and as the *Mercedita* only drew eighteen feet, there ought to have been no serious obstruction to her progress. It can be readily understood that the native pilots, who by long experience had become familiar with parts of the river more than ten and twelve feet in depth might yet be entirely ignorant of the limits of more restricted areas through which a vessel drawing eighteen feet could pass.

The first attempt ever made to map the channel of the Madeira below San Antonio was that of Messrs. R. H. Hepburn and George M. Keasby, with the steam tug *Brazil*, in 1878. These gentlemen, under instructions from Mackie, Scott, and Co., took many thousands of soundings and thoroughly examined every place in the river where navigation proved to be dangerous or difficult. The original map, made in two sheets by Mr. Keasby, is before the writer as he pens these lines. The scale was too small to show more than a few of the numerous soundings taken, but the deepest channel is definitely marked, while many marginal notes indicate the difficulties encountered and the best methods of avoiding them. Strong current, snags, exposed and submerged rocks, and sandbars are all clearly shown. The survey was particularly valuable because it was made between August 18 and September 6, when the water in the Madeira was rapidly approaching its lowest stage. Going over these maps today, we find only five places in the channel below San Antonio where the men on the *Brazil* found less than twelve feet of water. These are located as follows:

1. at the foot of Uroa Island, 11 feet;
2. at the mouth of the Rio Manicoré, 11 feet;
3. at the foot of Marmelos Island, 9 feet;
4. opposite Boca de Lago, 11 feet;
5. above Papagaios Island, 10 feet.

Commander Selfridge was sent to Brazil to map the channel of the Amazon and Madeira to San Antonio, but the results of his labors in 1878 were not made public until long after our return and consequently were of no use to us. He was but thirty-seven days on the Madeira and only succeeded in getting his vessel, the *Enterprise*, about one-third of the distance from the mouth of San Antonio. He never saw the river at or near either its maximum or minimum stage. He could easily determine from watermarks on trees the flood level of the river, but there was no possible way in which he could ascertain the annual rise and fall except by observations on water gauges established at various points for at least six months. The short time he was in the country made any such thing impossible, and the essential element required to reduce his soundings to low water must have been obtained from gossip with Indian pilots or ignorant natives on shore. This flaw vitiates his whole work and makes it unreliable except for a stage of water in the river at or near the one at which his soundings were made. He was at San Antonio only a few hours, yet, in that time, ascertained what none of the engineers there had yet been able to determine—that the annual rise and fall of the river below the falls was fifty-one feet. As a matter of fact, long continued and careful observations subsequently proved 41.75 feet to be the correct difference between high and low water. The consequence was that the Selfridge map of San Antonio Harbor shows depths of water less by 9.25 feet than really existed. Where the map indicates five feet of water on the bar below the steamboat landing, the same soundings, correctly reduced to low water, would give 14.25 feet, quite a serious error when we consider that no steamboat could land at San Antonio without crossing this bar. Strange to say, on all five of the general charts of the Madeira below San Antonio, Commander Selfridge states the annual rise and fall of the river to be thirty-eight feet, differing thirteen feet from his own determination at San Antonio and nearly four feet from that made by the engineers of P. and T. Collins! The mere statement that there is a *uniform* rise and fall in such a river for a distance of 661 miles, without any observations to prove it, is in itself perfectly absurd and incredible.

Malarial Fevers

Few persons residing in northern latitudes have any adequate conception of the great influence malarial fever in its various forms has exerted upon the history of tropical and semitropical countries. We need not go outside the United States to find large areas of fertile territory and populous suburban places that have been rendered undesirable for human habitation by this disease. It has been found on the top of the Catskills, along the banks of many of our great rivers, and in close proximity to noted health resorts on our seacoast and has even depopulated some districts in our Western states.

Malarial fever vanquished and demoralized our army in Cuba, while yet engaged in rejoicing over the conquest of a more insignificant foe. It was troublesome during the first American occupation of Havana, had made its influence felt in the Philippines, and is likely to prove a far more serious obstacle to progress in constructing the Panama Canal than even the greatly dreaded yellow fever.

It has been largely instrumental in maintaining the commercial isolation of Bolivia and has obstructed immigration and retarded the development of natural resources throughout a large part of South America, not only in the lowlands, but among the peaks of the Andes at an elevation of ten thousand feet above the sea. England has long been engaged in a conflict with malarial fever in her tropical colonies, in India, at Hong Kong, and in the vicinity of the Suez Canal. Greece, with a population of 2,433,806, was credited with 960,048 cases of the same disease during the year 1905.

In Italy, two million cases and fifteen thousand deaths *per annum* have been attributed to this malady. Nearly five million deaths *per annum* are believed to have been due to malarial fever in India, where one-third of the whole foreign population has been affected by it in a single year. Africa has always been notoriously unhealthy, mainly on account of the prevalence of malarial fevers. The hospital admissions of British soldiers infected with malaria in Sierra Leone each year averaged 213 percent of the whole force. Among the French soldiers in Algeria, the average number of cases of malarial fever per annum varied between 65 and 213 percent of the whole force. No definite statement can now be made regarding the number of cases in Brazil among employees of P. and T. Collins, but one of the physicians attached to the expedition, who had traveled extensively in Africa, assured the writer that the Dark Continent need have no terrors for anyone who could stand the climate of the Upper Madeira.

Because malaria has been almost invariably associated with an exuberantly productive soil and has had the effect of excluding millions of men from vast regions capable of subsisting them almost without toil, it was inevitable that sooner or later, all the forces of modern science would be combined to suppress it.

No military conflict in the history of the human race has ever been waged for stakes of such magnitude as those involved in the battle of the microscope with malaria, initiated in the year 1880, when Dr. Laveran, a French army surgeon serving in Algeria, discovered that malaria was caused by multitudes of minute amoeboid parasites living in the blood corpuscles of the patient.

It would be foreign to the purpose of this article to relate the varying fortunes that attended the many distinguished investigators in the crusade thus auspiciously begun or even to mention by name many who contributed directly or indirectly to the establishment and verification of the theory of malarial infection now generally accepted by scientific men. The leading facts are these: Prior to Laveran's discovery, Crawford, in 1807, and Notts, in 1847, had suggested the possibility of mosquito infection. In 1883, King and Koch took up the same idea and advanced theories on the subject. In 1894, Laveran gave out a conjecture tending in the same direction, and, in the same year, Manson independently arrived at a theory that had much in common with the views expressed by his predecessors.

For our present knowledge of the life history of the malarial parasite, for the identification of the particular kind of mosquitoes carrying the infection, and for rigid proof of the theory first conceived by Koch, we are indebted to years of laborious investigation carried on by Major Ronald Ross while connected with the Indian Medical Service of Great Britain. Although Major Ross began his study of malaria in 1881, it was not until 1898 that his labors were crowned with complete success and he proved beyond doubt or question the truth of his theory of malarial infection. That theory cannot be stated better than by quoting his own words, as used in a lecture on "The Progress of Tropical Medicine" delivered in Liverpool January 12, 1905. In it, he says:

We now know every phase and stage of the parasite of malaria as it exists, both in man and in the mosquito; and moreover we have ascertained that some 18 species of mosquito belonging to the single group of *Anophelina* can convey the disease, and that other kinds of mosquitoes appear to be innocent. Moreover the disease has actually been produced experimentally, first in the Indian researches on birds in 1898, and then by the Italians, by Fernside, Ziemann, and lastly by Manson. In 1900 Manson made the most striking of these experiments when he succeeded in infecting in the heart of London three gentlemen, who volunteered for the purpose, by means of mosquitoes brought from Italy. We therefore now know fully that the disease is carried by these insects, which convey the parasites from patients to healthy persons. The fact that malarial fever is connected with stagnant terrestrial water is fully explained. It is not because the germ lives in that water, but because the carrier of the germ, the *Anopheles* mosquito, does so. Strangely enough, the innocent varieties of mosquito do not live so much in stagnant terrestrial water as in vessels of water standing around houses—a fact which further justifies the old law. It is often asked whether malarial fever cannot be produced in other ways than by the bite of mosquitoes. I can only say that no satisfactory evidence of this has yet been given; and moreover that, for zoological reasons, it is exceedingly unlikely that the parasite, which already possesses a sufficiently complex life-history, can exist in other localities besides its own proper hosts, man and mosquito. At the same time no good reasons have yet been given for supposing such another phase of existence.

This theory naturally led to the suggestion of preventive measures, which were the following:

(1) The use of mosquito nets and wire gauze screens
(2) The use of *punkahs* and fans
(3) The cinchonisation of all patients
(4) The segregation of all infected persons
(5) Measures looking to complete or partial extermination of mosquitoes, such as the drainage of swampy places, filling of pools, dragging the weeds and periodic oiling of all stagnant water.

Notwithstanding the fact that these preventive measures were tried in many localities with most gratifying success, the theory propounded did not appeal with much force to those not versed in biology and untrained in the use of the microscope. There were thousands of persons all over the world who had lived so long in malarious districts that they considered themselves more competent to form a correct opinion on the subject than those who passed their time in dissecting mosquitoes and reading learned disquisitions before scientific associates. The theory, conclusively demonstrated by Major Ross and verified by others, was very slow, therefore, in finding acceptance among those most vitally interested.

Unfortunately, there were many actual or alleged facts

that gave plausibility to the objections urged against it. Mosquitoes, of the kind credited with carrying the malarial infection, were admitted to be abundant where malarial fevers were unknown. Persons were frequently attacked for the first time by malarial fevers in places where they were unable to detect the presence of mosquitoes of any kind. Mosquitoes had been known to migrate from the mainland of our coast to islands fifteen miles at sea, yet on many of the great rivers of the United States and South America, steamboatmen, constantly passing through and stopping at malarious places, enjoyed a most remarkable immunity from attack. In many countries, such as those in the area drained by the Amazon, where mosquitoes could be found in almost every locality and nothing apparently restricted their movements, malarial fevers were generally prevalent only in districts of comparatively small extent.

The stress placed by advocates of the theory upon the criminality of the mosquito seemed to imply that nothing taken into the stomach could cause or accelerate attacks of malarial fever. This idea was directly opposed to the practical experience of thousands of persons. For at least fifty years, it had been known in the valley of the Mississippi that eating watermelon is so frequently and so quickly followed by malarial fever as to convince many that the two occurrences stood in the relation of cause and effect. That "There is a shake in every melon" has long been a common saying among the Negroes of the South. In parts of South America, the same idea prevails with regard to sugarcane and ripe pineapples. At San Antonio, it was frequently noticed that even a moderate use of cognac brought on attacks of malarial fever, and it was universally believed by the natives that strong coffee had a tendency to ward off such attacks. Sleeping in clothing moistened by perspiration during the day was regarded as dangerous, and exposure to sudden changes of temperature equally so. None of these facts seemed consistent with the mosquito theory of infection to careless thinkers, though all can be satisfactorily explained without doing violence to it.

It is to be noted, however, that Major Ross does *not* state that mosquitoes are the *only* means of infection, though he evidently believes they are. Neither does he consider it essential or possible to decide whether man or the mosquito was the original host of the malarial parasite. Right here is the only vulnerable point in the whole theory, and it affects only that part that relates to the source from which the mosquito obtains the parasite. If the infection is carried only from one human being to another and does not originate in other than human beings, we are apparently forced to admit:

(1) That any number of healthy men traversing a country absolutely uninhabited by other human beings could by no possibility become infected with malaria

(2) That since the infection of a healthy person depends upon previous contact of a mosquito with one already infected, the smaller the number of human inhabitants in a specified region, the less would be the danger to uninfected persons passing through it

(3) That in a sparsely populated country, instead

of fighting countless millions of mosquitoes, the simplest and most practicable method of extirpating malaria would be to carefully segregate every human inhabitant until one generation of mosquitoes had perished and until the malarial parasite had been destroyed in the blood of those infected with it

The only comment the writer would make on the propositions stated above is that the country bordering upon the Madeira River for 240 miles above San Antonio, for many years, closely approximated the condition of a region uninhabited by human beings. Yet from 1852, when Lieutenant Gibbon explored it, until 1879, he claimed to know of no one who passed through it without a visitation of malarial fever. Further, it must be evident to anyone who has followed the foregoing narrative that the engineers of P. and T. Collins, while living in camp and almost completely isolated from other human beings, enjoyed no immunity whatever from malarial infection. Whether this experience has any important bearing upon the theory advanced by Major Ross and others is a question the writer leaves to those better qualified than himself to decide. However that may be, it is certain that the men who next attempt to build the Madeira and Mamoré Railway will have a powerful ally in modern science and that sanitary measures, unknown in 1878, will greatly facilitate their labors, if they profit by the lessons of the past and avail themselves of recent discoveries in regard to malarial infection.

Index

A

Acanga Pirangas (Red Heads) 363
Acre, Independent State of 453
actual work performed 426
Almeyrim, Mountains of 110
Amazon River(s) vii, viii, ix, xii, 52, 174, 177, 302
    Navigation Company 96, 105, 106, 111, 170, 301, 427
    volume of water carried by 118
animals 4, 85, 118, 119, 132, 227–230, 232, 236, 237, 250, 264, 265, 281, 287, 315, 329–331, 339
Ankers, Captain J. H. 155
anthropophagous acquaintances 354
Antofagasta, Port of 6
ants 126, 215, 236, 237, 242, 245, 249, 264, 275, 276, 279, 283, 313, 326–328, 334, 367, 380
Arara Indians 117
Arauz, Don Ignacio 120, 241, 311
Argentine railway 9, 10, 24
Arica, Port of 5
Arnold, Governor Samuel G. 172
Asuncion 7, 9
Atacama, Desert of 5, 14
at headquarters again 294
Audiencia of Charcas 1
Aunxt, Horace 387

B

Bacon, Vice Chancellor 50
Bahia de Marajo 107
Barbados 82, 83, 86, 87, 89, 90, 133, 197, 202, 203, 206, 209, 210
Barker, Wharton 176
Barlow, S. L. M. 32
    Wrecking Company 146
battle with the ants 264
Beer, Julius 33
Belcher, Edward 311
Bellvou, chief engineer 192, 200
Benjamin, Judah P. 437
Bird, Charles M. 68, 241
Blackbeard's castle 198, 199
Bly, Alem 311, 329
Bold Beach 146
Bolivia vii, viii, ix, xvi, xvii, xx, 1–6, 9–21, 28–31, 34–37, 38–43, 45, 47, 48, 51, 64, 117, 171–175, 177, 178, 231, 244, 246, 247, 407, 410, 428, 432, 433, 437, 444–447, 453–455, 461
    legal contest with 47
    natural resources of 35
Bolivian concession 32, 49, 433, 434, 437, 441, 443, 445
    annulment of 433
    loan 33, 41, 433, 436, 442, 443, 445
    tobacco 4, 96, 285, 366
Borba 17, 117
Brassey, Thomas 29
Brazilian concession revoked 444
Breves 108, 115, 128, 129
Bridgetown 83, 84, 86, 89
Brigido, Señor 125
Brisbin, Charles B. 402

Brooks, Richard W., statement of 147
Brown, James T. 216, 222, 304
    Milton 304
Buchholz, Charles W. 68, 241, 255, 290, 297, 300, 301
Bueno, M. A. Pimenta 96
Buenos Aires 9, 24, 29, 36
Bustillo, Don Rafael 17
Byers, Joseph 68, 128, 234, 374

C

Cachoeira de Caldeirão do Inferno 283
camp life in a rubber forest 349
Cape Henlopen 79
Caripuna Indians 251
Castlenau, Count 8
Ceara 268, 382, 385, 390, 392–394, 398, 427, 428
Charlotte Amalie, Harbor of 194
Chicha 350
Chills and fever 265, 282, 287, 290, 291, 293, 304, 321, 323, 324, 331, 349, 378, 401
City of Richmond 136, 164, 167, 169, 212, 217, 220, 227, 238, 269, 395, 427, 428, 434
Clay, Mr. 17
Clement, F. H. 262
climate 4, 5, 15, 16, 18, 95, 116, 197, 215, 246, 295, 296, 301, 307, 310, 364, 372, 383, 385, 412, 428, 435, 450, 462
Coates, Dr. Isaac T. 166
Cobija, Port of 5
Concepcion del Morrinhos 243, 318
conflicting claims 9
Cottrell, Captain 181, 193, 200, 209
Cox, John S. 268
Cozzens, Daniel M. 142
Craig, Neville B. 311
Crato 17, 120
Creighton, George W. xx, 224, 345

Cuatro Ojos 38
Cunningham, William 438
Currituck Beach 133, 138, 142
    Light House Club 148

D

Daniels, Darwin H. 261
Day, Mr. 270, 402, 439
day's work 226, 274, 276, 329, 349, 376, 385
Daza, President 48
Dead Dog Station 284
De Kierzkowski, C. F. 37
Delario, John 304
Denkin, Captain 181, 188, 189, 201, 206, 208, 210, 211
D'Invilliers, Camille S. xx
disintegration and collapse 372
Domatila Galves 367
Dom Pedro II 52, 53, 446
Dougherty, James 270
    John B. 68

E

Edison's camp 127
El Cerrito 42, 50
Elder, Captain 267
elevation of water surface 456
English failure and its consequences 28
Enterprise, the xi, xvii, 1, 20, 28, 29, 31–33, 35, 36, 40, 45, 48, 50–52, 58, 62, 132–134, 166, 172, 176, 234, 295, 300, 302, 307, 373, 375, 390, 400, 413, 414, 417–419, 424, 430, 443, 444, 459
Estreito do Breves 108
Eustis, William S. 267
Evans, Robert B. 268
events at San Antonio 214, 259
    resumed 259

Exaltacion 10, 11, 17, 19, 39, 42
expenditures incurred 424
exploring tour 130

F

Falls of Girão 16, 250, 451
    Theotonio 16, 242, 273, 292, 295, 311, 312, 316, 317, 319, 392, 393, 394, 428
Fetterman, Thomas J. 267
financial situation 303
first locomotive 306
Fisher, Morton 32
Flies 43, 275, 280–283, 292, 325, 326, 332
Frias, President 48
Fry, Justice 437, 439, 440

G

game 4, 82, 129, 132, 179, 246, 270, 273, 295, 330, 331, 344, 365
Gertie 72, 73, 81
Gibbon, Lieutenant, Lardner xvii, 14
Gillis, Captain 146
Gowen, Franklin B. 58, 62, 136, 171, 172, 255, 290, 297, 300, 301, 375, 381, 424
Grant, President ix, 53, 171
Green, Captain John P. 174
    Edward 341
Guajará-merim 15, 16, 31, 452, 454
Gurupa 109

H

Haénke, Thaddeus 12
Hageman, John, Jr. 158
Hamilton, Hugh 57, 58
    agreement with 58
Hampton, Mr. 149
Harrison, Jerome 304
Harris, Richard Reader 433, 443
Hartranft, Professor S. C. xxii, 385
Hastings 87
Hatteras, Coast of 77
Hayden, Charles J. xxi, 235
Hepburn, Robert H. xxi, 403
Hepburn's log 183, 200
Herbert, Arthur P. 304
Hiestand, J. H. 267
Hirsh, McClellan 268
Hoff, Samuel 355
Hopkins, George 36
Howard, fireman 195
Huanchaca mines 4
Huff, Benjamin 259
Humaita 120
human footprints 331, 354

I

Indians 1, 19, 22, 23, 36, 102, 103, 110, 117, 118, 121, 130, 131, 235, 242, 251, 252, 254, 257, 260, 263, 264, 268, 270, 272–274, 276–278, 280, 283, 284, 286, 287, 290, 303, 304, 310–313, 315, 316, 318, 319–322, 324, 325, 329, 332, 333, 335–338, 341, 343–347, 350, 351, 354–357, 359, 364, 365, 367, 368, 370, 391, 392, 396, 398, 401, 427, 428
Indios barbaros 260
Israel, Samuel 129

J

Jaci Paraná 103, 244, 255, 267, 302–305, 318, 323, 334–337, 339, 341, 346, 347, 349, 351, 352, 357–360, 364, 370, 389, 401, 450
    camp on the 346, 370, 401
    north bank of 318

Jamary River 121, 260, 363
Jameson and Hamilton 62
    Colonel John J. xx
Jessell, Sir George 432
Johnson, R. E. 267, 307
Jones, William 148
Josephs and Barger 62
Jumas 120

K

Keasby, George M. 458
Kehoe, Hugh 267, 269
Keller, Franz 18, 20, 121
Keller's report 339, 358, 362
King, C. J. 400
    Charles F. xiv, xxi, 165, 267, 400
    John, killed by savages 354, 355
    Miss Maria Juanita 388

L

laborers, class of 220
La Concepcion del Morrinhos 318
Lafourcade, Edward 156
Lambard, Charles A. 32
La Mott, Charles 400
La Paz 2, 5, 6, 16, 29, 36, 42, 179, 244, 433, 454
Las Pedras 302, 335, 352, 359, 365, 450
last days on the *Jaci Paraná* 364
Launch, the 39, 42, 111, 128, 129, 133, 242, 244, 274, 415, 416, 420
Laveran, Dr. 462
Lawford, Mr. 201
laying rails 259
legation to Mexico 28
Levick, Messrs. 176
Light House Club 148
Lima, Captain 405, 407–409
litigation in England 432
Lockwood 179

Lopez, fall of 25
Lorenz, Fred 91, 98, 291, 400
Loreto 7
Lunt, Benjamin
    Bros. and Co. 135

M

Macacos 128, 129, 215, 234–236, 238, 261
Mackie, Charles Paul 172
    experience at Pará 403
    James S. 32
    Scott and Company, Limited 175, 178
Madeira and Mamoré Association xvi, xvii, xxi, 421
    Company, Ltd. 36
    Railway xix, xx, xxi, 15, 31, 35, 36, 37, 57, 58, 136, 170, 172, 174, 241, 303, 394, 418, 424, 432, 441, 447, 451, 453, 455, 468
Madeira River xvi, xx, xxii, 18, 116, 171, 246, 255, 404, 406, 445, 452, 456, 458, 468
    earliest voyage on 11
    falls of 7, 13, 19, 29, 34, 178, 407
    flood discharge 119
    navigation of 13, 458
    survey of 171
Maher, T.C. 269
malarial fevers 16, 266, 296, 389, 461, 462, 466
Mamoré River 10, 12, 15–17, 20, 38, 41, 177
    the steamer 38
Manáos 13, 102, 223, 284, 389, 399, 449–451
Manning, Thomas 311
Marajo Bay 111, 128, 413
Mathews, Edward D. 42

Mato Grosso, mines of 11
Mauris, Maurice 167
McCalmont Bros. and Co. 59, 61, 62, 375
    agreement with 61
McIlvaine, Rodman 352, 389
McLane, Captain 405
McLaughlin, Frank 62
Melgarejo, President 36
Mercado, Don Santos 120, 256
Mercedita, cargo of the
    departure of the 134
    voyage of the 90
Metropolis, account of sailing
    history of the 141, 157
    wreck of the xix, 135, 140, 157, 161, 164, 180, 220, 428, 434
Mojos Basin 7
Monroe, Captain 160
Moore, Charles L. 167, 301
    James T. 136
Morales, President 36, 38
Morrinhos 130, 242, 243, 259, 274, 318, 320, 341
Morris, Ernest 422
    departure of 422
Morsing, Dr. C. A. 447
mosquitoes 243, 253, 275, 313, 318, 319, 326, 328, 367, 380, 384, 463–468

N

National Bolivian Navigation Company 29, 30, 31, 34, 37, 60, 173, 174, 178, 434, 442
Natividad 367
Nevins, W. S. 400
Newton, Alfred W. 167
Nichols, Othniel F. xx, 241

number of persons sent from the United States 427

O

Obidos 115, 116
O'Connor, John P. xxi, 270
O'Hara, John 76
"Old Mike," 297, 393
origin of the enterprise 1
Orton, Professor 97, 104
Oruro 6, 7
Oyola, Don Pastor 243, 255, 284, 290, 305, 318, 324, 333, 335, 338, 342, 346, 351, 352, 397

P

Palheta, Francisco de Mello 11
Pampean Sea 7
Pará xix, xxi, 10–12, 14, 15, 19, 41, 69, 90, 92–95, 97, 99–102, 104–106, 108, 111, 115, 116, 119, 127, 128, 133, 135, 138, 169, 171, 174–177, 179, 180, 182, 187, 189, 197, 203, 205, 207, 210–213, 216, 218, 245, 284, 299, 301, 302, 305–307, 347, 350, 352, 368, 372, 373, 376, 380, 382, 383, 386, 391 396, 399, 402–411, 413, 415, 417, 419, 420, 423, 428, 438, 439, 454
    churches of 100
    Fort of 97
    laundries of 100
    market of 351
    River 92, 93, 108, 128, 211, 218, 408, 419
    situation at 373
    to San Antonio 104, 119
Paraiso 121, 247, 256, 258
Parentintins 231, 357–363, 365

Parrish, Dillwyn 57
Patterson, C. H. 374
　　J. C. 311
peccaries 264, 265, 315, 329, 331
Pennington, Dr. Jack 179, 412, 423
Peto, Betts, and Co. 35, 36
Pilcomayo River 8
Pinkas, Senhor Julio 447
Pond, Fred 405
Pororoca, or bore 420
Porto de Mataura 17
Porto Mollendo 5
Potosi, silver deposits of 2
Poverty Flat 264, 265, 270
Prainha 110
Preston, Cecil A. xxi, 270, 280, 303
　　diary of 280
Public Works Construction Company
　　37, 41, 43–45, 47–49, 60, 65,
　　124, 424, 425, 434, 435
Puño 5, 6

Q

Quevedo, General Quintin 17, 28, 32

R

Railway, Argentine 9
rainfall at San Antonio 457
rains 8, 332
Ray, A. W. 433
recent revival of Colonel Church's
　　project 446
results x, 44, 140, 307, 424, 425, 441,
　　451, 454, 459
Reynolds, William H. 32
R. H. Bruce 68
Rio de Janeiro 30, 42, 51–54, 102,
　　171, 212, 392, 417, 444, 448
Rio São Lourenço 7
Roach Line of Steamers 102
Rodgers, Mike 378

Ross, Leathom Earle 37, 41
　　Major Ronald 463
Rosstown 130, 242
Runk, John 68, 129, 302

S

San Antonio xx, xxi, 15–17, 19, 31,
　　38–43, 50, 51, 59, 60, 67, 94,
　　102, 104–106, 119, 122–124,
　　126–134, 136, 166–168,
　　170, 171, 174, 177, 178, 180,
　　212–216, 219–222, 225–227,
　　231–235, 240–242, 245, 247,
　　255, 256, 259–262, 265, 269,
　　270, 272, 277, 283, 284, 290,
　　292, 294, 295, 299, 300, 301–
　　307, 310–313, 316, 320, 332,
　　334, 338, 357, 358, 360, 361,
　　363, 364, 369, 370, 372, 374–
　　376, 379, 381, 382, 384–386,
　　388–390, 392, 394, 396, 398,
　　400–407, 409–412, 425–430,
　　434, 438, 439, 447, 449–452,
　　454, 457, 458–460, 466, 468
　　Harbor, map of 460
　　headquarters at 402
　　hospital 299
　　life at 225, 227
San Carlos 130, 216, 241, 242, 301,
　　302, 305, 316–320, 343, 358,
　　361, 370
San Fernando 24
San Patricio 310, 322–324, 329, 332,
　　333, 335–337, 341–343, 346,
　　355, 356, 370
Santa Cruz de la Sierra 19, 38
Santarem 114, 427
Santa Rosa 389
Sapucaia-Oroca 117, 119
Schele, George A. 304, 352, 359
Schooner *Eva I. Smith* 136
　　John S. Wood 136